THE ANATOMY *of*
INSECTS & SPIDERS

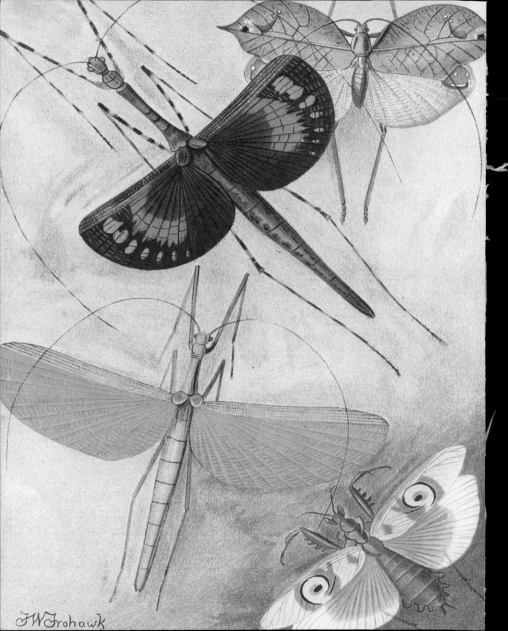

F.W.Frohawk

THE ANATOMY *of*
INSECTS & SPIDERS

Over 600 Exquisite Forms

By Claire Beverley & David Ponsonby

CHRONICLE BOOKS
SAN FRANCISCO

First published in the United States in 2003
by CHRONICLE BOOKS LLC

Copyright © 2003 THE IVY PRESS LIMITED.
*All rights reserved. No part of this book may be
reproduced in any form without written
permission from the publisher.*

*Library of Congress Cataloging-in-Publication
Data available.*

ISBN: *0-8118-3983-4*

Manufactured in CHINA
Cover design by AYAKO AKAZAWA

Distributed in Canada by RAINCOAST BOOKS
*9050 Shaughnessy Street
Vancouver, British Columbia V6P 6E5*

10 9 8 7 6 5 4 3 2 1

CHRONICLE BOOKS LLC
85 Second Street, San Francisco, CA 94105

www.chroniclebooks.com

This book was created by
THE IVY PRESS LTD.
The Old Candlemakers,
Lewes, East Sussex BN7 2NZ

Creative Director PETER BRIDGEWATER

Publisher SOPHIE COLLINS

Editorial Director STEVE LUCK

Design Manager TONY SEDDON

Designer JANE LANAWAY

Senior Project Editor REBECCA SARACENO

Copy Editor MANDY GREENFIELD

DTP Designer CHRIS LANAWAY

Picture Researcher LORRAINE HARRISON

CONTENTS

INTRODUCTION

I NSECTS AND SPIDERS *have always been important to humans: as a food source; as vital pollinators or protectors of our crops; as providers of fabrics to clothe us and dyes; as medicines; as a threat to our well-being; as mystic symbols of immortality, power, doom, or good luck; and simply as animals to enjoy and study. The oldest records of our relationship with them date from 9,000-year-old cave paintings in Spain. However, the founding of the modern science of entomology is credited to the philosopher and scientist Aristotle, who in the 4th century BC named insects and spiders "Entoma," meaning "the notched ones." Aristotle's influence was to dominate entomology for the next millennium. It was another thousand years before the next land-mark, and during that period, known as the Dark Ages, the Church dominated scientific thinking. Then, in AD 1255 the monk Albertus*

Magnus published the first of a 26-volume work, De Animalibus. *He succeeded in breaking the grip of theology on science, paving the way for the Renaissance in the 15th century. However, it was not until Francesco Redi made use of the microscope in the 1680s that modern experimental biology was formed and the first accurate accounts of insect metamorphosis were published. By the end of the next century, Carolus Linnaeus and Johann Christian Fabricius had described many species. But it was the French zoologist Pierre-André Latreille, in his work of 1796, who laid the foundations of modern entomology.*

The great voyages by naturalists such as Alexander von Humboldt and Charles Darwin, the expeditions to South America in the 1840s by Alfred Wallace and Henry Bates, and the pioneering work of John Le Conte and Thomas Say in North America in the 19th century—all undoubtedly popularized entomology and helped people comprehend the diversity among this wonderful group of creatures.

TAXONOMIC TERMS

Taxonomy is the theory and practice of classifying living things into closely related groups that are known as *taxa* (singular: *taxon*). The Greek philosopher Plato (ca. 429–347 BC) was the first to provide written definitions of abstract classification terms: "Each individuum is only an imperfect reproduction of a perfect eternal conception of the species and genus." However, it was his pupil Aristotle who produced the first classification system based on appearance (anatomy) and function (physiology). His system may have been crude and incomplete by current standards, but it was to endure for more than 2,000 years.

Today taxonomy forms a part of a wider process known as systematics. Systematists classify organisms using a wide range of characteristics that help to explain how a particular group has evolved, including fossil evidence and DNA analysis. However, its foundations

Charles Darwin's voyages to South America on the *Beagle* were instrumental in arousing a wider interest in entomology.

predate Charles Darwin's theories on the origins of species and thus, as with Aristotle's system, it was originally based on anatomy, physiology, and life cycles.

Many scientists contributed to the new system of nomenclature, but it was the Swede Carl von Linné (1707–78) (known to many as Carolus Linnaeus) who justly gave his name to the structure that is now in use: the Linnaean System.

Examples of the hierarchy that is used to provide each organism with its own unique classification in the Linnaean System are given in the table opposite. Explanations for the various taxonomic terms may be found in the Glossary (*pages 276–9*).

Carolus Linnaeus's outstanding contribution to science was his system of nomenclature for plants and animals, as outlined in his many publications.

LINNAEAN CLASSIFICATION

Biologists still argue over the definition of a species, but under Linneaus's system, each is given two names, one identifying the species itself, the other, the genus to which the species belong. Today, most of us use the common names. Below is a table with two examples of classification, one for the hornet, *Vespa crabro*, and the other for the desert locust, *Schistocerca gregaria*:

Taxon	Hornet	Desert locust
KINGDOM	Animalia	Animalia
PHYLUM	Arthropoda	Arthropoda
CLASS	Insecta	Insecta
SUB-CLASS	Pterygota	Pterygota
DIVISION	Endopterygota	Exopterygota
ORDER	Hymenoptera	Orthoptera
SUPERFAMILY	Vespidoidea	Acridoidea
FAMILY	Vespidae	Acrididae
GENUS	*Vespa*	*Schistocerca*
SPECIES	*crabro*	*gregaria*

ABOUT THIS BOOK

THIS BOOK *is an introduction to the immense—and immensely complex—world of insects. The aim in compiling a selection of beautiful line engravings, some of them more than 200 years old, is to give the reader a glimpse of the sheer variety within this world and to offer some history, anecdotes, and strange facts that surround the subject. This approach means the book cannot be used as a technical introduction or as a field guide. We hope it is a visual pleasure and a satisfying read; and if it engages you sufficiently to send you in search of more "serious" reference works, then it will have done its job (for some suggested modern works, see page 280).*

Because all the sources used were very old, a policy for the naming of the different insects had to be formulated. Much of the taxonomy has changed since the source material was written—

reclassification is perpetual in a constantly developing field—so we realized that by using old engravings, the taxonomy could not be consistent or, in some cases, accurate. Hence, where the insects were not named in the original, we have tried, especially if integral to the text, to supply further identification. Also, if identification was incorrect, where possible it has been corrected. A ✎ symbol is placed against the name where further or corrected identification is provided, so it is clear this naming is contemporary. A misspelled word has been rectified without adding the ✎ symbol. If, however, the engraving prohibited accuracy beyond doubt, the original labelling was left. For help with the careful and proficient identification of the images, we would like to give special thanks to Mike Darton. Finally, please be aware that reference to the use of insects or spiders as medicines should be interpreted in the context of the anecdotes and not taken as recommendation for modern use. ✎

ANATOMY OF AN INSECT

Insects possess bodies with hard external skeletons organized into small compartments called segments. Segments vary in number and organization but they are arranged into three distinct sections, the head (typically consisting of 6 segments at the embryo stage), the thorax (3 segments), and the abdomen (10 or 11 segments). Each segment has a specialized function and may also have a pair of appendages that can be used for various activities, including sensing, eating, walking, flying, or reproduction.

001	Vertex
002	Lateral lobe of pronotum
003	Coxa
004	Trochanter
005	Spine
006	Spur
007	Tarsus
008	Tegmen
009	Tibia
010	Knee
011	Hind femur
012	Pronotum

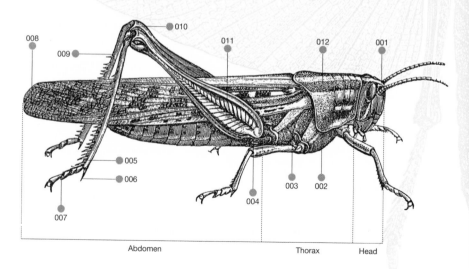

Abdomen   Thorax   Head

001 Antenna	008 Supraanal plate	015 Wing
002 Prozona	009 Cercus	016 Tegmen
003 Lateral carina	010 Abdomen	017 Costal field
004 Metazona	011 Disc	018 Costal margin
005 Femur	012 Spur	019 Band
006 Tibia	013 Band	020 Sulcus
007 Tarsus	014 Apex	021 Median carina

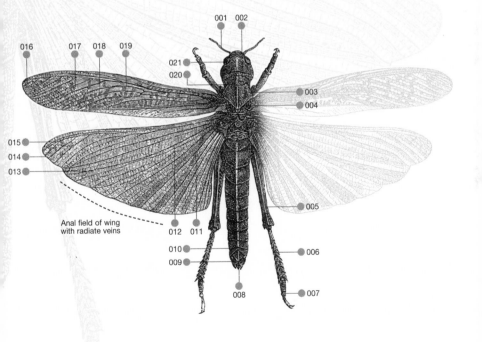

Anal field of wing
with radiate veins

BEETLES *Coleoptera*

B eetles form a group of more than 350,000 known species. The order name of Coleoptera given to this captivating group is derived from the Greek names for sheath (*coleos*) and wing (*pteron*). Beetles come in a fascinating range of body shapes and alluring metallic sheens, vibrant colors and creative markings. Perhaps the more infamous members are those beetles that present themselves as pests, such as the destructive larder beetles and the pestiferous chafers—the bane of the unfortunate plant lover who well understands the decimation that they cause. However, there are other beetles that have gained popularity as a result of their attractive colors and markings, together with their valuable predatory behavior against insect pests—we are, of course, referring to the ladybugs. Then there are the Hercules and Goliath beetles of the tropics: titans of the beetle group, weighing in at 3½ oz/100 g.

As with some of the more esthetically pleasing insects (due to their color, shape, or size), 19th-century literature explains how these colossal beetles fetched high prices among collectors in the past. Equally fashionable as collectors' pieces were the marvellously shaped *Mormolyce phylloides* and the superb harlequin longhorn beetles. Glass-fronted display cabinets bursting with such treasures were not the only places for more unusual specimens; indeed decorations in ladies' hair, brooches, and necklaces were but a few of the ways in which they were exploited. In some cultures entomophagy (the consumption of insects) developed, and it remains inherent among Africans, Aboriginal Australians, and Native Americans. Larvae of certain beetles are prepared with a variety of ingredients, the result being a good source of nourishment. The Coleoptera order of insects are indeed a mesmerizing group to behold.

Class	Insecta
Sub-Class	Pterygota
Division	Endopterygota
Order	

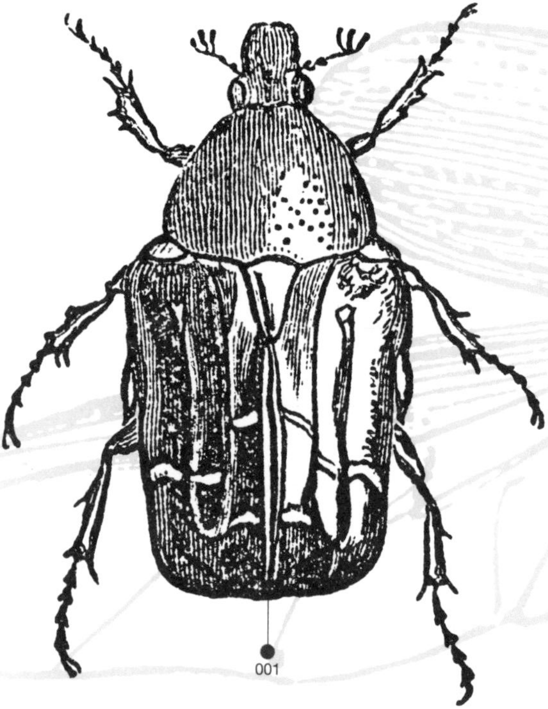

001

002

003

004

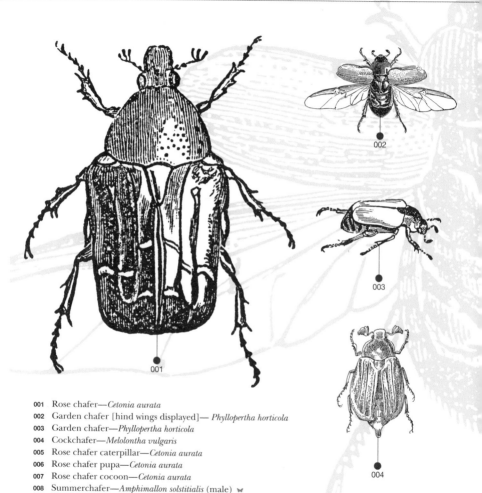

001 Rose chafer—*Cetonia aurata*
002 Garden chafer [hind wings displayed]— *Phylloperthа horticola*
003 Garden chafer—*Phyllopertha horticola*
004 Cockchafer—*Melolontha vulgaris*
005 Rose chafer caterpillar—*Cetonia aurata*
006 Rose chafer pupa—*Cetonia aurata*
007 Rose chafer cocoon—*Cetonia aurata*
008 Summerchafer—*Amphimallon solstitialis* (male)

005

006

007

Chafers, although majestic in appearance, with their characteristic fanlike (lamellate) antennae, are in a group of Coleoptera of agricultural importance: the scarab beetles. Their pestiferous lifestyle now somewhat overshadows any esthetic qualities they may have, although in the past rose chafers were given homes of bamboo cages and became pets for ladies in Manila. The genus name of *Melolontha* is derived from Greek words meaning "apple tree" and a "flowering" or "inflorescence." Larvae of the cockchafer beetle (commonly referred to as the Maybug or June beetle, *see no. 004*) were used by the Swedes to predict the severity of winters, based on their color. And in England it was thought that the Maybug larvae gave rise to briars.

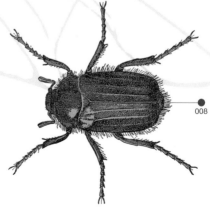

008

Class	INSECTA		
Sub-Class		PTERYGOTA	
Division			ENDOPTERYGOTA
Order			

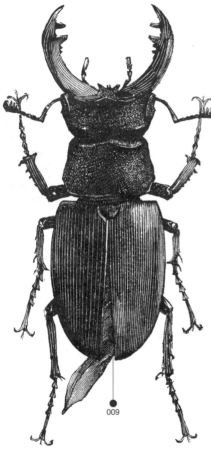

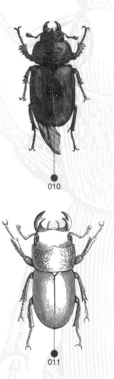

009   Stag beetle—*Lucanus cervus* (male) ⋈
010   Stag beetle—*Lucanus cervus* (female) ⋈
011   Lesser stag beetle—*Dorcus parallelipipedus* ⋈
012   Dungpushers
013   Rhinoceros beetle
014   Common rhinoceros beetle
015   Common Typhaeus—*Typhaeus fumatus*

012

The common name of stag beetle (*no. 009*) originates from the imposing antler-like mouthparts (mandibles) of the male. But despite their impressive size, it is actually the smaller jaws of the female that administer the stronger pinch (*no. 010*). According to the American entomologist Frank Cowan, the ancients named elephants and oxen *lucas, lucana*, thereby arousing speculation that this was the origin of the stag beetle's family name, Lucanidae. Another theory is that it was based on a group of Italians called the Lucanians. The beetles simply named dungpushers (*no. 012*) are relations of the stag beetles, and are also known as tumblebugs, referring to their movement when transferring dung. One particular species of dung beetle is dubbed the "lousy watchman" in reference to its mite infestation.

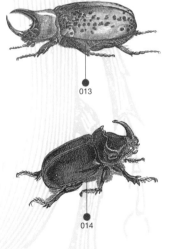
013

014

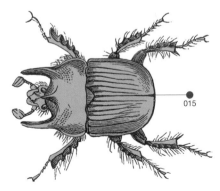
015

Class	INSECTA		
Sub-Class		PTERYGOTA	
Division			ENDOPTERYGOTA
Order			

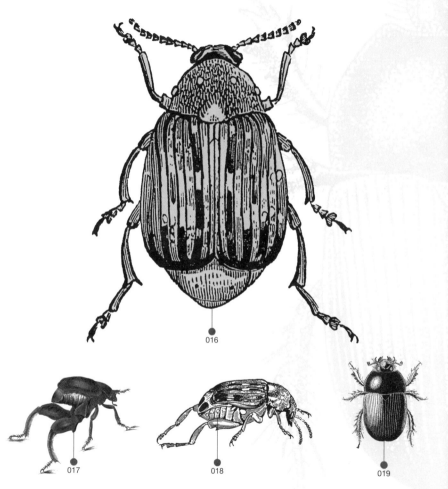

016

017

018

019

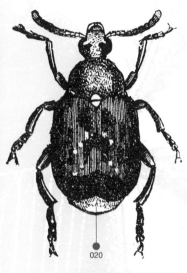

020

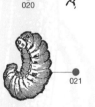

021

The superfamily of Scarabaeoidea was known as the Lamellicornia in older literature, a word that describes their lamellate (fanlike) antennae. In 19th-century literature the scarab beetle depicted here (*no. 017*) was aptly named the kangaroo beetle for its well-developed hind legs. And in ancient Egypt, scarabs, or sacred beetles, were worshiped as symbols of resurrection and immortality. However, the larvae pose a problem as serious pests of plant roots. Clearly depicted in the side-view illustration of the bean beetle (*no. 018*) is a slight extension of the head, although these seedeaters do not have true extended snouts (rostrums), as weevils do. Their hardened forewings (elytra) do not entirely cover the abdomen. This is another distinctive feature of these seedeating pests.

016 Bean beetle—*Acanthoscelides granarius* ✤
017 Scarab beetle—*Lucanus cervus* ✤
018 Bean beetle—*Acanthoscelides flavimanus* ✤
019 Scarab beetle—*Lucanus cervus* ✤
020 Red footed weevil—*Bruchus rufimanus*
021 Red footed weevil larva
022 Shiny leaf beetle—
      Chrysophora chrysochlora ✤

● 022

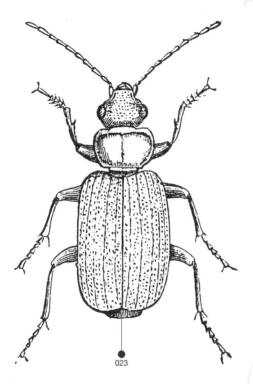

023

024

025

026

023 Tiger beetle—*Lebia cyanocephala*
024 Tiger beetle—*Notiophilus biguttatus*
025 *Anchomenus dorsalis*
026 Violet ground beetle—*Carabus violaceus*
027 *Amara obsoleta*
028 Violet ground beetle
029 Bombardier beetle—*Brachinus crepitans* ◄

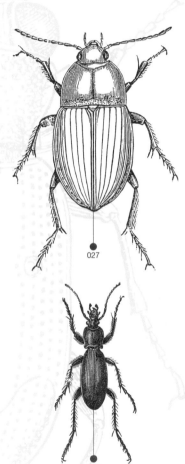

027

Ground beetles were known as Geodephaga by early entomologists, meaning "earth eaters" or "devourers of the earth." This was a somewhat misleading common name for adults and larvae that are fundamentally carnivorous, although the intention seems to have been to distinguish these terrestrial beetles from the Hydradephaga, meaning "devourers of the water," in reference to the aquatic beetles. Some Carabidae, or ground beetles, lack hindwings and their forewings are fused together. The beautiful bluish-purple tinge to the edges of the violet ground beetle's elytra (*no. 026*) earned it its common name, while the bombardier beetle (*no. 029*) emits a puff of a volatile chemical, followed by a "popping" sound, thereby giving it its name. Early naturalists likened this form of defense to humans' use of gunfire.

● 029

028

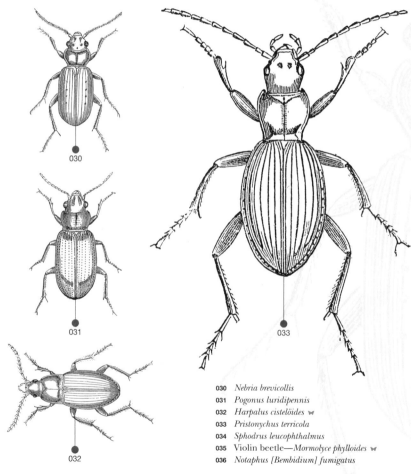

030 *Nebria brevicollis*
031 *Pogonus luridipennis*
032 *Harpalus cistelöides* ✆
033 *Pristonychus terricola*
034 *Sphodrus leucophthalmus*
035 Violin beetle—*Mormolyce phylloides* ✆
036 *Notaphus [Bembidium] fumigatus*

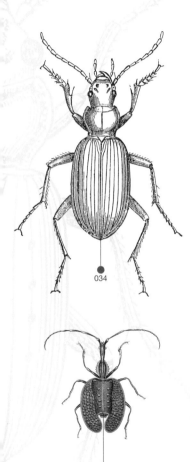

034

For beetles that spend their time either on or under the ground, the common name of ground beetle is accurately descriptive. The predatory Carabidae have long been recognized as successful biological control agents of insect pests. However, a perhaps less obvious use for this group was developed in some parts of Africa, where it is documented that the native people prepared soap from one large carabid. Clearly shown on this page are the longitudinal ridges of the elytra (which are commonly black, but are sometimes clothed in an alluring metallic luster) and the easily recognizable body shape of these beetles. However, such a distinctive body shape is by no means the only form in this group. The magnificently shaped violin beetle (*no. 035*) is a native of Java and represents yet another eye-catching specimen, which fetched a high price among collectors of old.

036

035

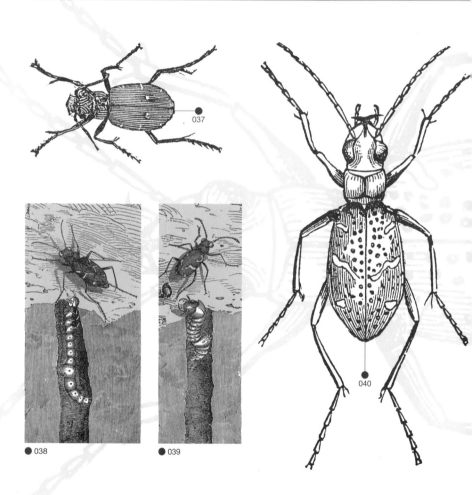

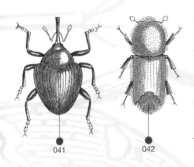

041   042

Carolus Linnaeus assigned the genus name *Cicindela* to members of the Cicindelidae family, a term that refers to the destructive and luminous insects. The name "tiger beetle" (*see no.s 037–040*) conjures up a rather menacing beetle with large mandibles, and this group of beautiful beetles does indeed contain fierce predators with strong jaws. With their rapid flight and fast-running nature as adults and the larval habit of burying themselves in the ground awaiting unsuspecting prey, the justification for their common name becomes apparent. The larvae are able to reside in burrows because of their humped backs—their heads being camouflaged in the entrances of the tunnels. Meanwhile, bark beetles are perhaps best known through the spread of the devastating Dutch elm disease.

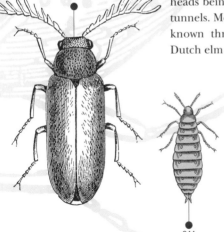

043

044

037 Green tiger beetle—*Cicindela campestris* ✎
038 Wood tiger beetle—*Cicindela sylvatica* (and larva)
039 Dune tiger beetle—*Cicindela maritima* (and larva) ✎
040 Wood tiger beetle—*Cicindela sylvatica* ✎
041 *Orobites cyaneus*
042 *Ips typographicus* ✎
043 *Drilus flavescens* (male)
044 *Drilus flavescens* (female)

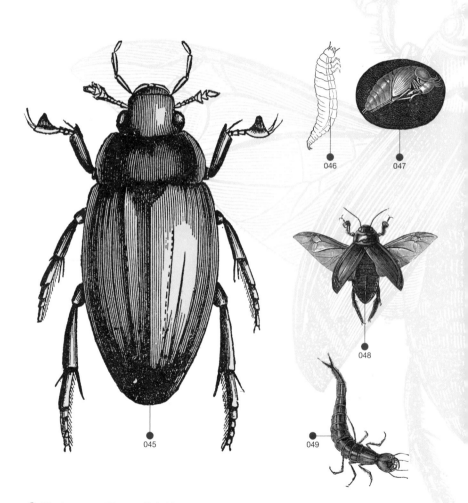

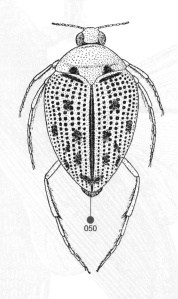

050

The robust form of the larger water beetles is quite distinctive. The great water beetle, also called the great silver beetle (*no. 045*), possesses clubbed antennae, with the ends of the structures enlarged to form their characteristic shape. Measuring approximately half the width of a floppy disk, this aquatic beetle is one of Europe's largest, while another water-frequenting beetle, the beautiful, teardrop-shaped haliplid beetle (*no. 050*) crawls in aquatic plants in shallow water. Depicted here are the fearsome jaws of the larva (*no. 049*), whose use earned them the French name *ver-assassin*—implying an affinity for blood and destruction. The adult great silver beetle (*no. 051*) suspends itself in an inverted position at the water's surface to replenish a large air store concealed below the wingcases.

045 Great silver beetle—*Hydrophilus piceus* (male) ✷
046 Great silver beetle larva ✷
047 Great silver beetle pupa ✷
048 Great silver beetle—*Hydrophilus piceus* ✷
049 Great water beetle larva
050 Haliplid water beetle
051 Great silver beetle—*Hydrophilus piceus* ✷

051

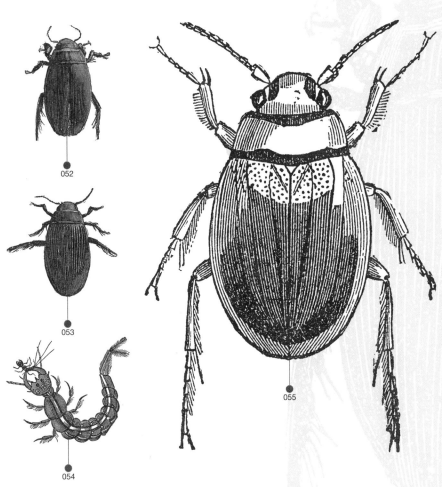

052

053

054

055

052 Great diving beetle—*Dytiscus marginalis* (male) ✎
053 Great diving beetle—*Dytiscus marginalis* (female) ✎
054 *Dytiscus* larva
055 Screech beetle—*Hygrobia hermanni* ✎
056 *Dytiscus* pupa
057 *Agabus biguttatus*
058 Whirligig beetle—*Gyrinus natator*
059 Whirligig beetle larva—*Gyrinus natator*

The diving beetle's genus name of *Dytiscus* was first applied by Carolus Linnaeus and aptly originates from the Greek word for diver. Marginally apparent in the illustration of the male (*no. 052*) are the slightly wider forelegs, compared with the female of the species (*no. 053*); these are used to hold the female during mating. As for the petite and gregarious whirligig beetle (*no. 058*), both its common name and its genus name of *Gyrinus* are derived from the Greek for "move in a circle," pertaining to its whirling movement on the water's surface. The ingenuity of these beetles' eyesight is astounding. Their compound eyes have an upper and lower part, assisting them to see both above and below the water.

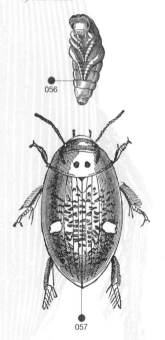

056

057

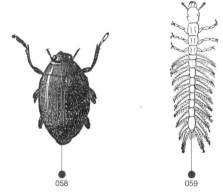

058

059

Class	INSECTA
Sub-Class	PTERYGOTA
Division	ENDOPTERYGOTA
Order	

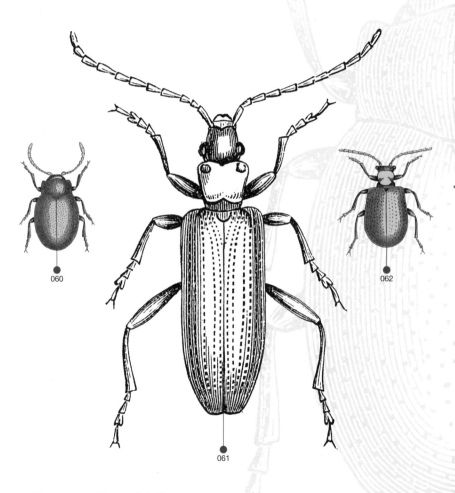

060

061

062

Chrysomelidae, the family name of leaf-eating beetles (*no. 060*), denotes "an apple of gold." Indeed, it was this beautiful golden color that seduced Chilean and Brazilian ladies to fashion necklaces from them; and in western America they were used to decorate boxes. Some members of this family frequent aquatic environments where their larvae feed on aquatic plants. Another associate of this family is the bloody-nosed beetle (*no. 063*), so called because of its tendency to reflex-bleed; when disturbed, vivid red fluid exudes from its mouth and joints. The larvae of one handsome leaf-eating species of Chrysomelidae live in "pots" constructed from the beetles' own feces—a fascinating habit that has earned them the common name of pot beetles (*no. 064*). The adults sport stunning, burnished orange elytra.

063

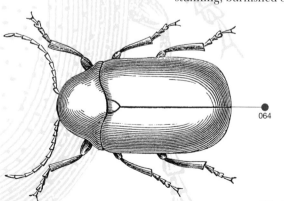

064

> *"Whenever I hear of the capture of rare beetles, I feel like an old war-horse at the sound of a trumpet."*
> CHARLES DARWIN

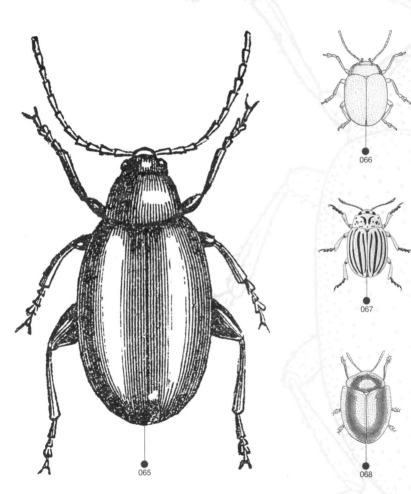

065

066

067

068

> *"It seems…that a taste for collecting beetles is some indication of future success in life."*
> CHARLES DARWIN

The notorious Colorado beetle (*no. 067*), thought to have originated in Mexico, is a critical pest of the potato family. It was accidentally transported to Europe in the early 20th century and spread rapidly through potato crops, causing widespread devastation. The American entomologist Thomas Say first described specimens from the Rocky Mountains in the early 1800s. Colorado beetles are highly distinctive in appearance, with longitudinal black stripes on a yellow wing case. The tortoise beetle (*no. 068*), with its oval shape and flattened sides, bears some resemblance to the creature of the same name. Herbivorous and beautifully colored in some cases, these tiny beetles made exquisite jewelry pieces in the past.

065 Turnip flea beetle—*Phyllotreta brassicae* ✎
066 Leaf beetle—*Chrysolina staphylea*
067 Colorado beetle—*Leptinotarsa decemlineata* ✎
068 Green tortoise beetle—*Cassida viridis*
069 Asparagus beetle—*Crioceris asparagi*
070 Turnip flea beetle—*Phyllotreta nemorum*

069

070

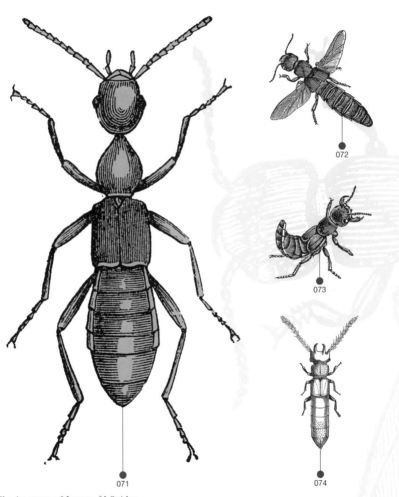

071

072

073

074

COLEOPTERA

BEETLES 11

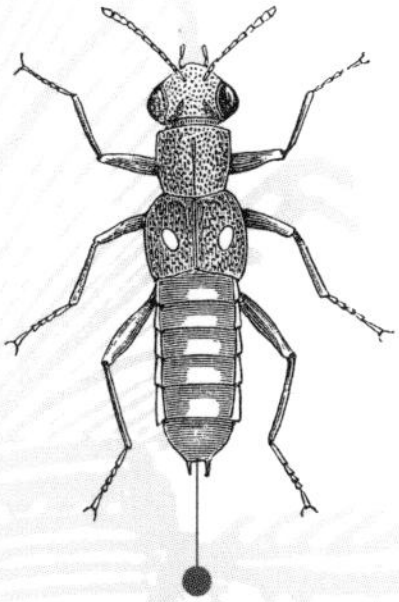

075

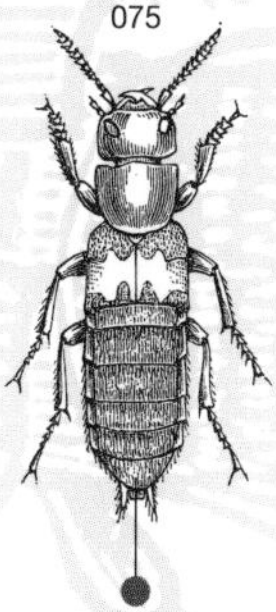

076

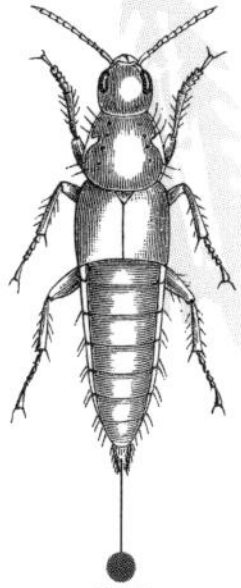

077

One immediately obvious features of all rove beetles is their shortened elytra, which do not entirely cover the abdomen. The older name of Brachelytra for this group was derived from the Greek *brachys*, meaning "short"—a reference to this characteristic. Linnaeus included all these beetles in the genus *Staphylinus*, but there are now known to be in the region of 3,500 European species in a multitude of genera, of which just a few are represented here. Clearly shown in the illustration of the devil's coach-horse beetle or the fetid rove beetle (*no. 072*), in disturbed stance, are the rather ferocious-looking jaws, which are used in attacking prey. Also noticeable are the bulging eyes of a rove beetle (*no. 075*), which usually lives near an aquatic environment.

071 Red-necked rove beetle
072 Devil's coach-horse or fetid rove beetle—*Ocypus olens* (full-grown)
073 Devil's coach-horse or fetid rove beetle larva—*Ocypus olens*
074 Four-horned rove beetle
075 *Stenus bimaculatus*
076 Hairy rove beetle—*Creophilus maxillosus*
077 *Philonthus marginatus*

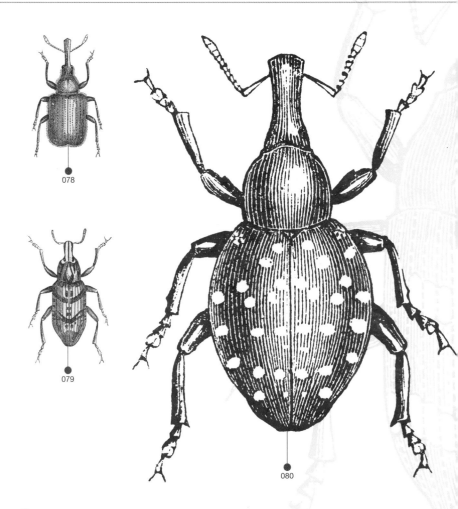

078

079

080

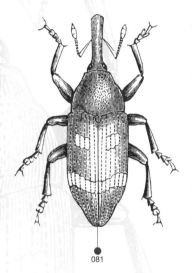

081

The ancient name of Rhynchophora for the group of beetles known today as weevils is more descriptive than their modern name, for *rhynchos* is Greek for "snout." Indeed, the distinguishing long proboscis is undoubtedly the first feature to be noticed on these beetles, to which Linnaeus assigned the genus name of *Curculio.* An array of interesting anecdotes, ranging from fashion to lawsuits, surrounds weevils: For instance, we discover from a 19th-century source that not only were the grubs eaten, but it was believed that they were responsible for milk production in women. Other fanciful accounts include a poem that describes weevil larvae developing into butterflies before being transformed into mice. And a Florentine publication of the 18th century reported that a cure for toothache was to be found in a specific weevil.

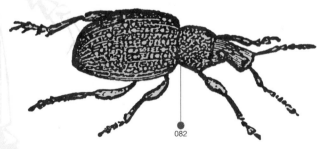

078 *Rhynchites bacchus*
079 *Cleonus nebulosus*
080 *Moltytes germanus*
081 Pine weevil—*Pissodes pini*
082 Black vine weevil—
   Otiorhynchus sulcatus

082

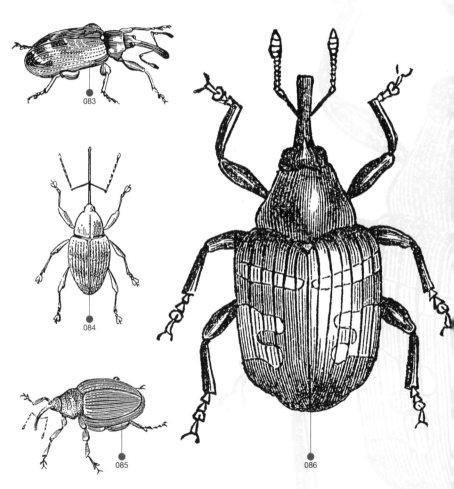

083

084

085

086

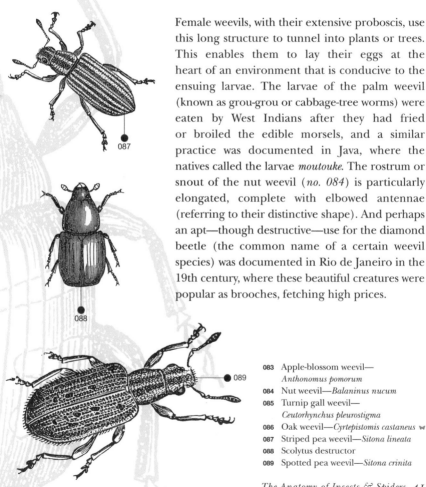

Female weevils, with their extensive proboscis, use this long structure to tunnel into plants or trees. This enables them to lay their eggs at the heart of an environment that is conducive to the ensuing larvae. The larvae of the palm weevil (known as grou-grou or cabbage-tree worms) were eaten by West Indians after they had fried or broiled the edible morsels, and a similar practice was documented in Java, where the natives called the larvae *moutouke.* The rostrum or snout of the nut weevil (*no. 084*) is particularly elongated, complete with elbowed antennae (referring to their distinctive shape). And perhaps an apt—though destructive—use for the diamond beetle (the common name of a certain weevil species) was documented in Rio de Janeiro in the 19th century, where these beautiful creatures were popular as brooches, fetching high prices.

083 Apple-blossom weevil—
    Anthonomus pomorum
084 Nut weevil—*Balaninus nucum*
085 Turnip gall weevil—
    Ceutorhynchus pleurostigma
086 Oak weevil—*Cyrtepistomis castaneus* ⌖
087 Striped pea weevil—*Sitona lineata*
088 Scolytus destructor
089 Spotted pea weevil—*Sitona crinita*

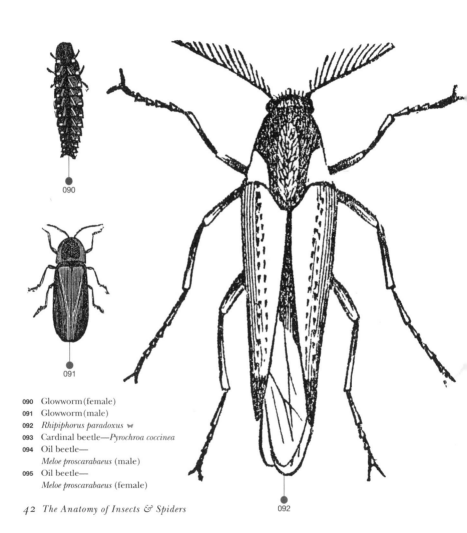

090 Glowworm (female)
091 Glowworm (male)
092 *Rhipiphorus paradoxus* ✎
093 Cardinal beetle—*Pyrochroa coccinea*
094 Oil beetle—
   Meloe proscarabaeus (male)
095 Oil beetle—
   Meloe proscarabaeus (female)

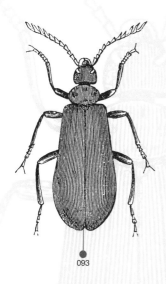

094

Rhipiphorus paradoxus (*no. 092*) was so named because people were, in the past, puzzled about its habits. Another enigmatic beetle is the cardinal beetle (*no. 093*), which, in contrast to the bark under which it hides, has a glorious red blush. The larvae of oil beetles (*no. 094*)—a relation of the cardinal beetle—crawl from their eggs to flowers, from where they are carried on the bee to its nest; here they prey upon bee eggs, pollen, and nectar. Another dazzling specimen in this collection of images is the glowworm, the female of which is wingless and does indeed appear more like a worm than a beetle (*no. 090*). Interesting anecdotes surrounding glowworms include them being mistaken for troubled spirits. A popular use for these radiant insects, acknowledged in both India and Italy, was as a decoration for ladies' hair.

094

095

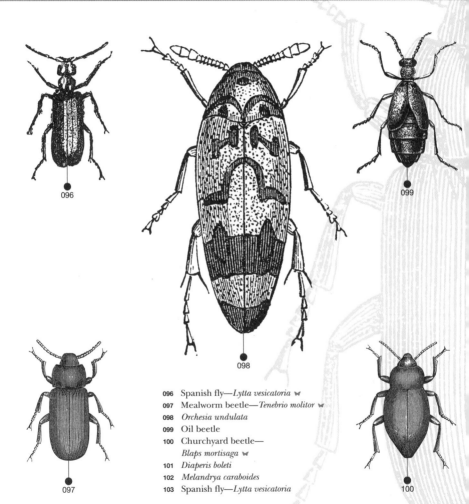

096 Spanish fly—*Lytta vesicatoria* ✎
097 Mealworm beetle—*Tenebrio molitor* ✎
098 *Orchesia undulata*
099 Oil beetle
100 Churchyard beetle—
　　Blaps mortisaga ✎
101 *Diaperis boleti*
102 *Melandrya caraboides*
103 Spanish fly—*Lytta vesicatoria*

The term *mortisaga*—the old species name of the flightless churchyard beetle (*no. 100*)—came about (or so the story goes) because of its alarming appearance and the fact that it was labeled a harbinger of death. In contrast to modern views pertaining to diet, Egyptian, Arabian, and Turkish women prepared this beetle with butter, spices, honey, and other ingredients and then ate it with the singular desire of gaining portly figures. Other interesting uses for beetles are seen in the blister beetle or Spanish fly (*no. 103*). Not only did apothecaries of old use this as a source of the remedy cantharides, but they prescribed it as an antidote against earache and scorpion "bites"— notwithstanding its reputation as an aphrodisiac.

101

102

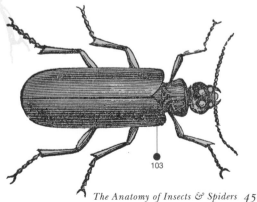

103

Class	INSECTA	
Sub-Class		PTERYGOTA
Division		ENDOPTERYGOTA
Order		

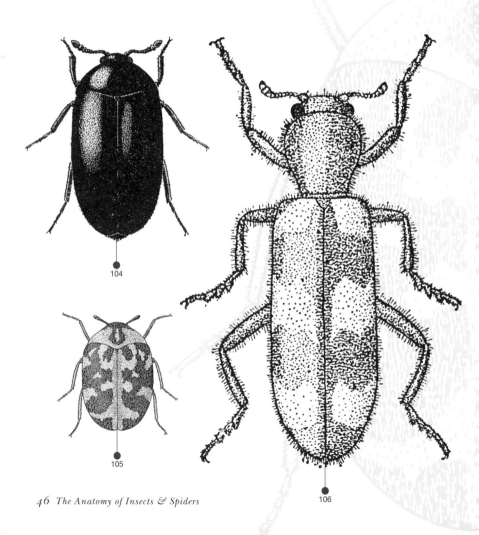

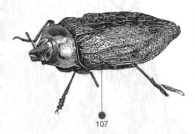

107

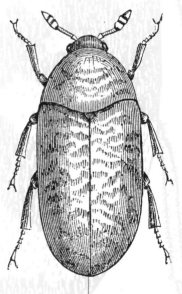

108

Carpet beetles (*no. 104 or 105*) attack rugs, carpets, clothing, and furniture. The larvae are very hairy and are commonly known as "woolly bears." These beetles share the family name Dermestidae with larder beetles, which are employed to clean animal skeletons. In 1865 Frank Cowan retold an interesting account of beetles of this family being found in a stone coffin in the "Links of Skail"(exactly where he meant is unknown) and in a mummy's head from Thebes. It was surmised that the burying of these beetles with the bodies represented some form of communication. The Larder beetles' (*no. 108*) destructive lifestyle has led to them becoming renowned as pests of hides and dried meat; one pestiferous species prompted an offer of a £20,000 reward to anyone who could solve the problem of damage to animal skins in London warehouses.

104 Black carpet beetle
105 Buffalo carpet beetle—*Anthrenus scrophulariae* ➤
106 Clerid beetle—*Trichodes scrophulariae* ➤
107 Species of splendor beetle—*Euchroma gigantea*
108 Larder beetle—*Dermestes murinus* ➤

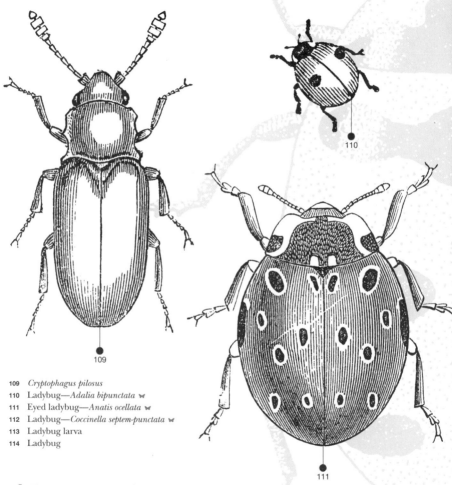

109 *Cryptophagus pilosus*
110 Ladybug—*Adalia bipunctata* ⋙
111 Eyed ladybug—*Anatis ocellata* ⋙
112 Ladybug—*Coccinella septem-punctata* ⋙
113 Ladybug larva
114 Ladybug

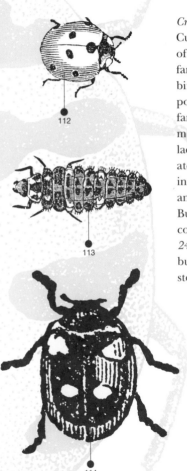

112

113

114

Cryptophagus beetles are from the superfamily Cucujoidea and feed mainly on fungus. But the often colorful Coccinellidae of the same superfamily are perhaps better known—ladybugs, ladybirds, lady beetles, or lady cows have become popular models for modern merchandise. The family name is derived from the Latin *coccinatus*, meaning "clothed in scarlet." However, not all ladybugs are red; indeed, not all are even punctuated with spots. Ancient documented practice includes the use of ladybugs as palliatives for colic and measles, and to relieve the pain of toothache. But following early use of the vedalia ladybug to control the cottony cushion scale insect (*see page 246*) in California in 1888, the importance of ladybugs as a means of biological control has withstood the test of time and remains with us today.

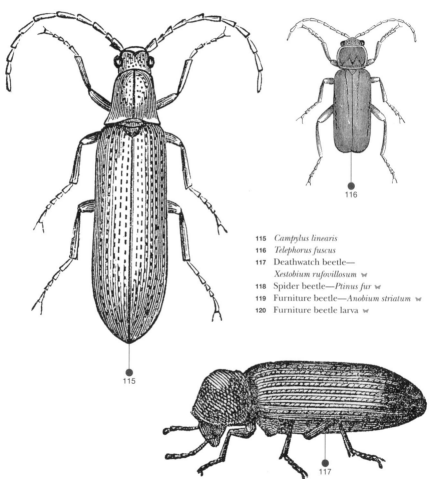

115 *Campylus linearis*
116 *Telephorus fuscus*
117 Deathwatch beetle—
   Xestobium rufovillosum ✎
118 Spider beetle—*Ptinus fur* ✎
119 Furniture beetle—*Anobium striatum* ✎
120 Furniture beetle larva ✎

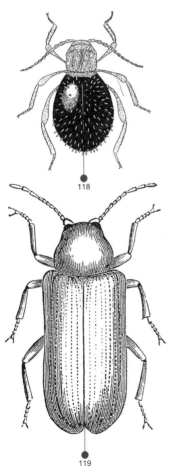

118

119

The sinister-sounding common name of the deathwatch beetle (*no. 117*) derives from the fact that this notorious pest of wood bangs its head against the wood to make its presence known, thereby producing a knocking noise, like the sound of a watch ticking; a superstition associated with this banging states that it foretells misfortune or death. The ambiguously named spider beetles (*no. 118*) are undoubtedly members of the Coleoptera—it is the long legs and hairy bodies of these insects that earn them their spider-like associations. As for *Campylus linearis* (*no. 115*), the first word means "curved" in Greek, essentially the opposite of the straightness suggested by *linearis*.

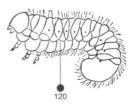

120

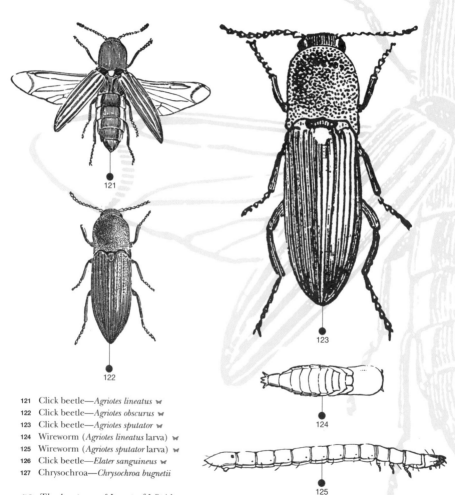

121 Click beetle—*Agriotes lineatus* ❧
122 Click beetle—*Agriotes obscurus* ❧
123 Click beetle—*Agriotes sputator* ❧
124 Wireworm (*Agriotes lineatus* larva) ❧
125 Wireworm (*Agriotes sputator* larva) ❧
126 Click beetle—*Elater sanguineus* ❧
127 Chrysochroa—*Chrysochroa bugnetii*

126

The larvae of the family Elateridae are commonly referred to as wireworms (*no. 124*) because they are stiff and do not bend easily. They may live in this juvenile state for three to five years before developing into slender adults. Beautifully depicted with its elytra open to reveal the veined hind-wings, this beetle is commonly called "click beetle" (*see no.s 121–123 or 126*) or "skipjack," referring to the noise made when it flicks itself into the air, to right itself if it has fallen onto its back. The resulting clicking action offers a shock feature against predation, as well as letting the beetle regain its footing. The adult beetles feed onpollen and nectar, but the larvae are serious crop pests.

127

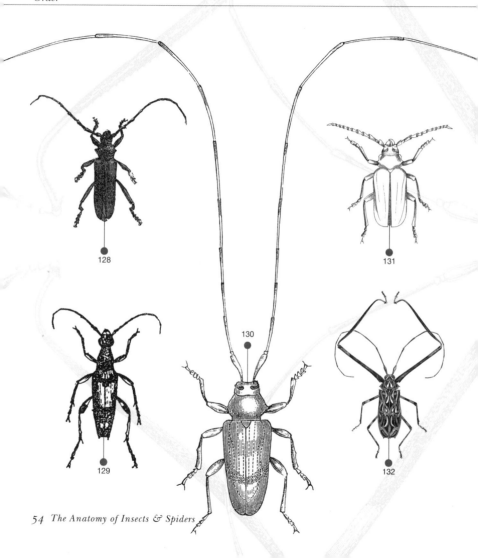

In ancient Egypt the markings on the exquisite harlequin longhorn or harlequin beetle (*no. 132*) were thought to be hieroglyphics. And it was known by natives of tropical parts of America as *Mouche bagasse*—the name being derived from a tree with bright yellow wood, which exudes a strong-smelling liquid that was used by early collectors to attract the insect. One collector of this beauty was the scientist Sir Joseph Banks on the voyage of the *Endeavour* to the South Pacific in the latter part of the 18th century. And one South American member of the longhorn family warrants special mention due to the length of its body: slightly smaller than the height of this book!

● 133

● 134

ANTS, WASPS, BEES, & SAWFLIES *Hymenoptera*

Ants, wasps (including hornets), bees, and sawflies all belong to a group of insects that Aristotle named Hymenoptera—from *hymen*, meaning "membrane." The word is particularly appropriate because it not only refers to the beautiful, membranous wings that all but worker ants in this group possess, but also has connections to Hymen, the Greek god of marriage—by virtue of the fact that the fore- and hindwings are joined together by little hooks. To date, more than 100,000 species have been described, but some entomologists believe that many more are yet to be discovered.

Since the Renaissance most entomologists have divided these insects into two groups: broad-waisted sawflies (considered to have evolved earliest) and wasp-waisted bees, wasps, and ants. The latter are the only insects to have evolved a sting from their egg-laying organs (ovipositors) and the only group, other than termites, to have developed complex social behavior. They attracted the attention of humans earlier than many other insect groups. For example, the harvesting of honey from bees was first recorded 9,000 years ago in a rock painting in the Cuevas de la Araña in Valencia, Spain, and since then ancient and modern cultures have studied them in great detail. Indeed, one of the first detailed graphic records of any insect was that of a hornet, dating from 3100 BC in ancient Egypt. So accurate was this picture that it was even possible to identify the species.

A vegetarian diet is common among sawflies, gall wasps, bees, and some ants, whereas most wasps are predatory or parasitic. Large hunting wasps have always attracted the attention of humans, but the Victorians were staggered to learn of the diversity among parasitic wasps, and discovered thousands of species. We now know that many invertebrates (including most insects) are host to at least one species of parasitic wasp that lays its eggs in the body or eggs of its victim.

Class	INSECTA		
Sub-Class		PTERYGOTA	
Division			ENDOPTERYGOTA
Order			

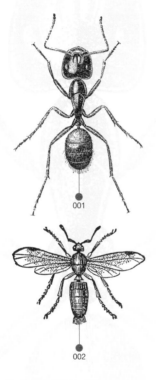

001

002

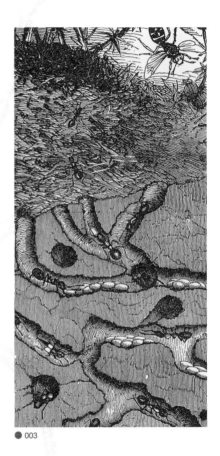

● 003

001   Unidentified ant ⩗
002   Jet ant—*Formica fuliginosa* ⩗
003   Portion of the nest of the red ant
004   Pupa of ant
005   Larva of ant
006   Black ant—*Formica fuliginosa*
007   Black ant—*Formica fuliginosa*

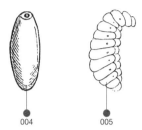

004

005

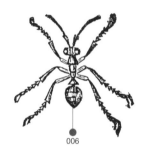

006

Many of the post-Renaissance entomologists were amateurs who combined their hobby with their profession. Clergymen in particular contributed much to this branch of science, and many missionaries came back from expeditions into the unexplored parts of the various continents with tales of spectacular insect phenomena and much valuable information. One of these missionaries from the Congo tells of a narrow escape when, lying sick in his bed, he was rescued by his African servants from a great marching army of black ants, 3 miles/4.8 km long and 6 feet/1.8 m wide, that surged through his mission. The ants ate up every living thing, including an unfortunate cow left in one of the stables. Only its bones were found.

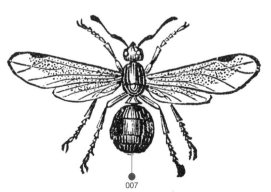

007

"A drop of amber from the weeping plant,
fell unexpected and embalmed an Ant;
The little insect we so much contemn
Is, from a worthless Ant, become a gem."
M. V. MARTIAL

008   Section of an anthill
009   Red wood ant (worker)—*Formica rufa* 🪰
010   Red wood ant (male)
011   Unidentified ant 🪰
012   Red wood ant (female)

● 008

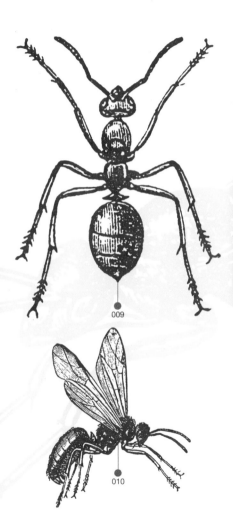

009

010

011

"Turn on the prudent Ant thy heedless eyes, Observe her labors, sluggard! and be wise."

SAMUEL JOHNSON

The American entomologist Frank Cowan, in his 1865 book *Curious Facts in the History of Insects*, states, "The laborious life and foresight of the ant have been celebrated throughout antiquity, and from the wise Solomon down to the amiable La Fontaine, the sluggard has been referred to this insect to 'Learn her ways and be wise'" (Proverbs 6:6). Thus was the unselfish labor of the worker ants, which live only to serve their queen and colony, viewed by the industrious and patriotic Victorians. The ancient Greeks shared this belief, as demonstrated by the fable of the Myrmidons who, like ants, applied themselves to diligent labor and hard toiling and thereby became rich.

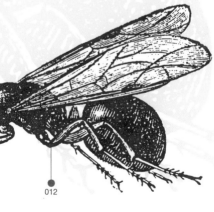

012

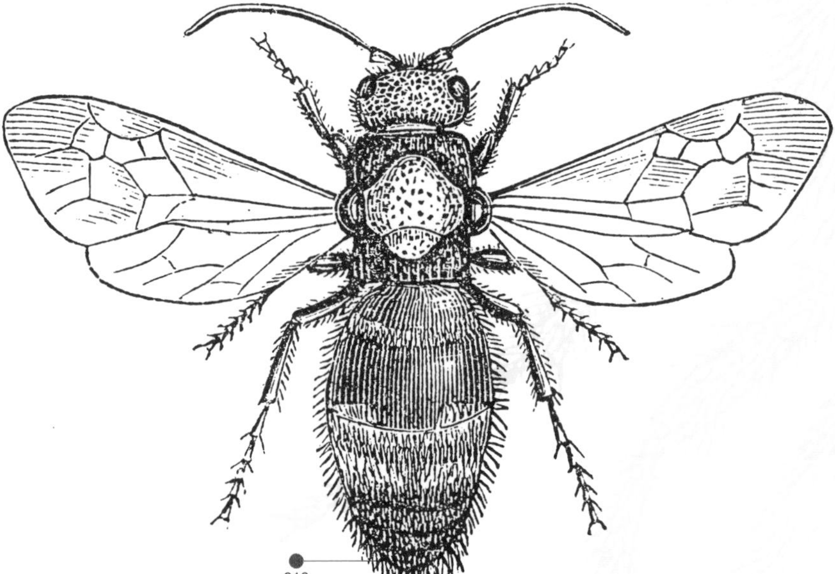

013

- **013** *Mutilla europaea* (male)
- **014** *Mutilla europaea* (female)
- **015** Velvet ant—*Mutilla* sp
- **016** Leaf cutting ant (male)—*Oecodoma cephalotes*
- **017** Wood ant (female)
- **018** Leaf cutting ant (winged female)—*Oecodoma cephalotes*
- **019** Wood ant (worker)
- **020** Wood ant (male)

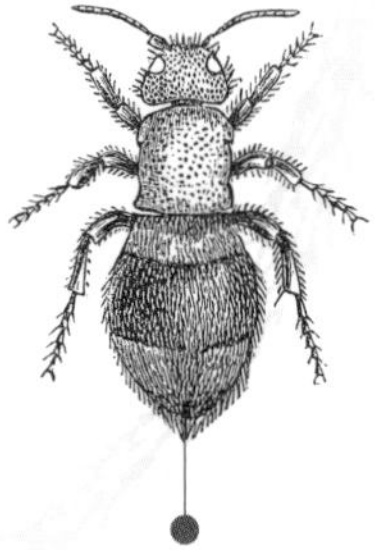

014

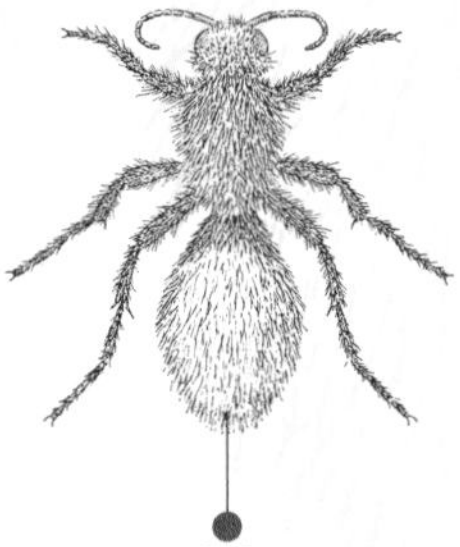

015

ANTS &
VELVET ANTS 3

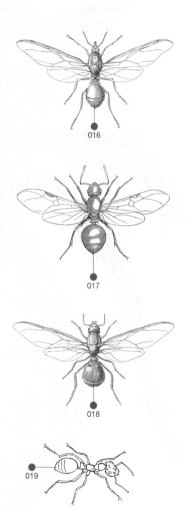

016

017

018

019

Two of the earliest examples of the use of natural enemies to control insect pests both involve ants. In 3rd-century China, ant nests were sold near Canton to control pests on citrus trees; and the Talmud (Mo'ed Katan 6b) describes how ants from one nest could be used to destroy a rival ants' nest. What is remarkable about the latter account is that it shows how the early Hebrew entomologists had a clear knowledge of ant trails (where worker ants from one colony leave a chemical trail for their sisters to follow) and accurately recognized how ants disperse in order to set up new colonies. Modern entomologists have only rediscovered this information in the last 50 years.

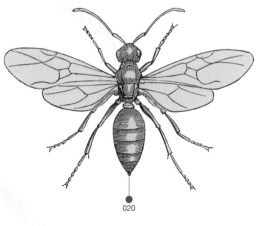

020

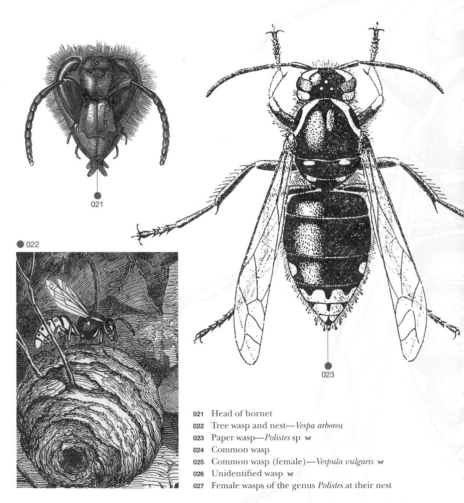

021   Head of hornet
022   Tree wasp and nest—*Vespa arborea*
023   Paper wasp—*Polistes* sp ✎
024   Common wasp
025   Common wasp (female)—*Vespula vulgaris* ✎
026   Unidentified wasp ✎
027   Female wasps of the genus *Polistes* at their nest

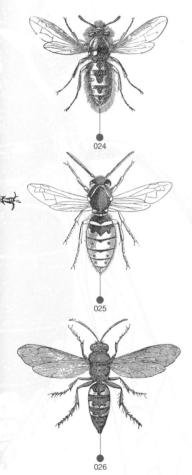

024

025

026

Most ants do not sting (although they may inflict bites and squirt formic acid at their enemies). But "true" wasps certainly do have painful stings, and many (like the ants) form a highly developed social organization. Also like ants, wasp colonies consist of sterile workers, males, and a queen who founds the colony and lays the eggs. The larvae are housed in large, paper nests made from chewed wood. The workers are expert hunters, feeding their larval charges on masticated insects. In America such wasps are known as yellow jackets, and Victorian entomologists delighted in extracting half-built nests to study their behavior. The ancient Greek and Roman entomologists believed that wasps were "sprung from the dead bodies of horses."

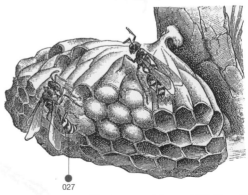

027

The Anatomy of Insects & Spiders 65

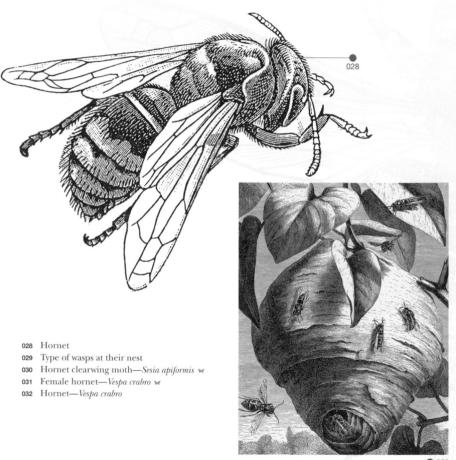

028

028 Hornet
029 Type of wasps at their nest
030 Hornet clearwing moth—*Sesia apiformis* ✎
031 Female hornet—*Vespa crabro* ✎
032 Hornet—*Vespa crabro*

● 029

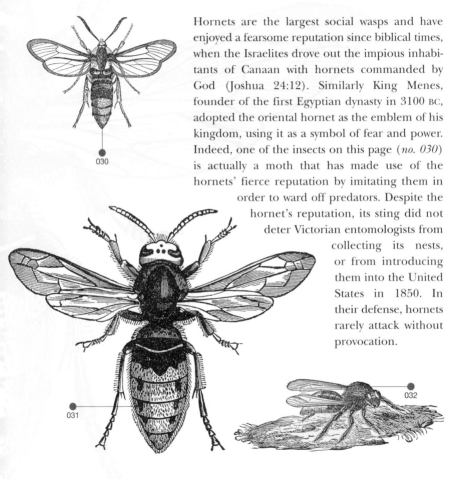

030

Hornets are the largest social wasps and have enjoyed a fearsome reputation since biblical times, when the Israelites drove out the impious inhabitants of Canaan with hornets commanded by God (Joshua 24:12). Similarly King Menes, founder of the first Egyptian dynasty in 3100 BC, adopted the oriental hornet as the emblem of his kingdom, using it as a symbol of fear and power. Indeed, one of the insects on this page (*no. 030*) is actually a moth that has made use of the hornets' fierce reputation by imitating them in order to ward off predators. Despite the hornet's reputation, its sting did not deter Victorian entomologists from collecting its nests, or from introducing them into the United States in 1850. In their defense, hornets rarely attack without provocation.

032

031

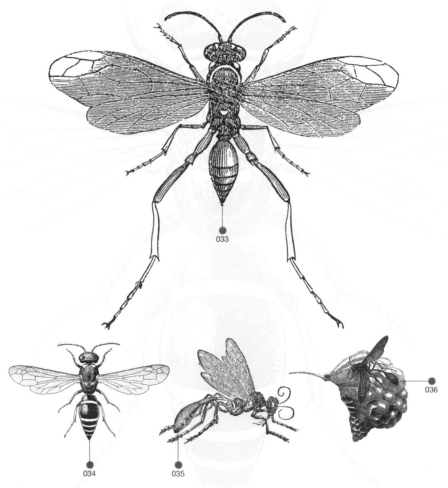

033

034

035

036

037

Solitary wasps (*no. 038*) differ from other parasitic species in that they build a nest in which to house their eggs and the paralyzed prey upon which the emerging larva feeds. Potter wasps (*no. 034*) fashion nests out of clay, making vaselike structures, while others simply dig a tunnel in sandy places or tunnel into hollow plant stems. Then they catch their unfortunate prey—a spider, caterpillar, fly, or beetle—and paralyze it by stinging then lay an egg in it. Early entomologists in India and China, while observing the entombment of the caterpillar by these wasps, did not see her lay her egg and believed that she sang a magical song as she sealed up the nest, thus starting the gradual transformation from caterpillar to wasp.

033   Unidentified wasp
034   Potter wasp—Eumenidae
035   Sand wasp
036   Unidentified wasp
037   Spider wasp—Pompilidae
038   Solitary wasp and nest—*Eumenes* sp
039   Scoliid wasp—*Scolia flavifrons*

039

● 038

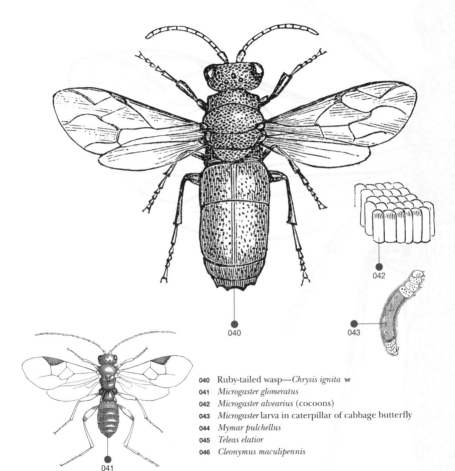

040 Ruby-tailed wasp—*Chrysis ignita* ✎
041 *Microgaster glomeratus*
042 *Microgaster alvearius* (cocoons)
043 *Microgaster* larva in caterpillar of cabbage butterfly
044 *Mymar pulchellus*
045 *Teleas elatior*
046 *Cleonymus maculipennis*

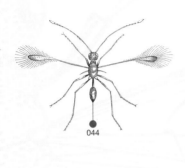

044

Parasitic wasps are mostly tiny creatures less than ⅛ inch/3 mm in length. The fairy flies from the family Mymaridae are the smallest of all insects, measuring less than 1/100 inch/0.25 mm. They lay their tiny eggs in the eggs of other insects. Other species variously parasitize the larva, pupa, nymph, or adult of the host, using their ovipositor to lay eggs directly into their body tissues. The victim is then slowly killed as the wasp larva hatches and feeds on the tissues and body fluids. Late Victorian economic entomologists (pest controllers) studied these creatures with interest and went on daring expeditions in unexplored countries to discover new species that preyed on agricultural pests.

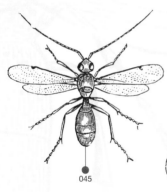

045

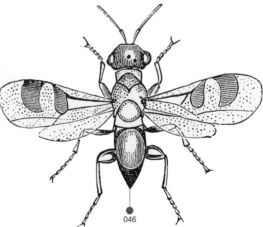

046

Class	INSECTA		
Sub-Class		PTERYGOTA	
Division			ENDOPTERYGOTA
Order			

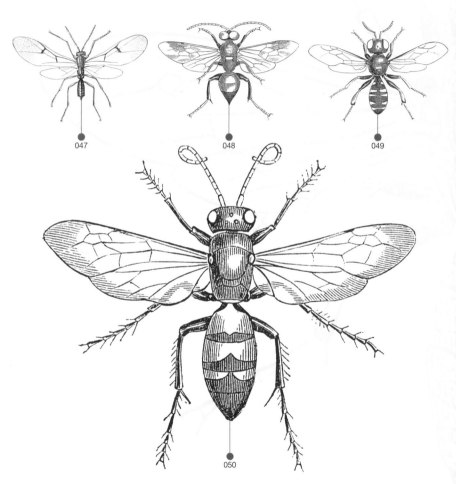

047

048

049

050

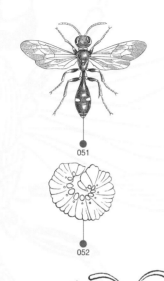

051

052

Most of the wasps on this page are solitary spider-hunting wasps. During the 16th and 17th centuries American colonists in the southern states were amazed to find several extremely large species, up to 1½ inches/37 mm long, that parasitized tarantulas. But many of the adults, while being expert hunters and providing meat for their offspring, feed entirely on nectar. Close relatives of these and the parasitic wasps are the gall flies. These have developed relationships with plants so that when the gall flies lay their eggs in the plant, chemicals injected with them cause the plant to grow into a home with food for the hatching larvae. Examples of such abnormal growths are the familiar oak apple and robin's pincushion.

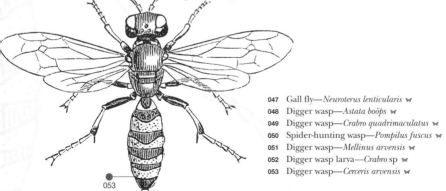

053

047 Gall fly—*Neuroterus lenticularis* ✎
048 Digger wasp—*Astata boöps* ✎
049 Digger wasp—*Crabro quadrimaculatus* ✎
050 Spider-hunting wasp—*Pompilus fuscus* ✎
051 Digger wasp—*Mellinus arvensis* ✎
052 Digger wasp larva—*Crabro* sp ✎
053 Digger wasp—*Cerceris arvensis* ✎

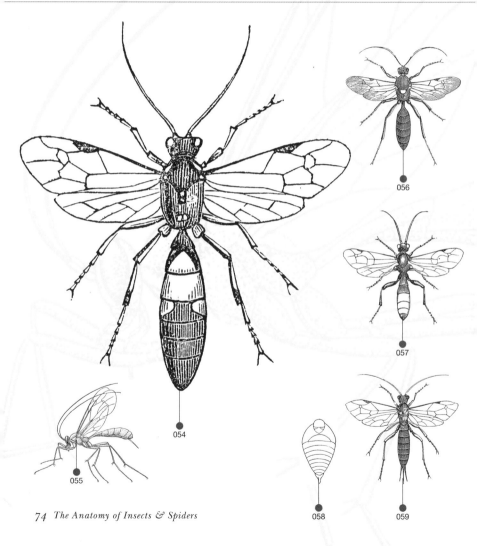

054

055

056

057

058

059

060

061

062

Ichneumon flies and their close relatives, the braconids, are mostly parasites of butterfly and moth caterpillars. One of their victims is the silkworm and, since this has been in cultivation for at least 6,700 years, it is likely that the ancient Chinese were well aware of these parasites. They and the Victorian entomologists of the Western world had certainly learned much about their life history by the end of the 19th century. For example, L. Trouvelot, writing in the *American Naturalist* in 1867, described in fascinating detail the way in which the female ichneumonid carefully deposits her eggs on the caterpillar and how the larvae first feed on the fatty tissues, before switching to the vital organs and killing their host just after it forms a cocoon.

054 *Ichneumon crassorius*
055 Ichneumonid wasp ⩗
056 *Ichneumon proteus*
057 *Tryphon rutilator*
058 *Tryphon* larva
059 *Pimpla instigator*
060 *Cryptus migrator*
061 Ichneumonid wasp ⩗
062 Ichneumonid wasp ⩗

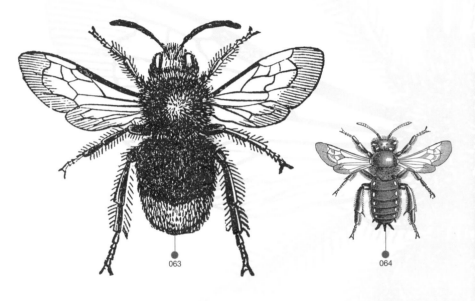

063

064

065

063 Stone humble bee (male)
064 Hoop shaver bee
065 Stone humble bee (female)
066 Stone humble bee (worker)
067 Cells from the nest of the common humble bee
068 Shaver bee

Humble bees (or bumblebees) are amazing creatures. Physical laws suggest that their shape and weight should preclude them from flying, but in midsummer they may be on the wing for 20 or more hours each day. The Victorians viewed their untiring labor as an inspiring example to the "sluggard." Humble bees are important pollinators and entomologists have now succeeded in domesticating them to use for this purpose in greenhouses. In the wild, as Beatrix Potter's Mrs. Tittlemouse found out, the queen often colonizes deserted mouse holes. There she will raise a small brood of worker females and, later in the year, males. Queens live for only one season, so the nests are abandoned each year.

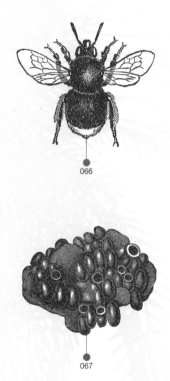

066

067

068

069

070

● 071

069 Humble bee 🦋
070 Humble bee 🦋
071 Moss humble bee and nest
072 Humble bee 🦋
073 Humble bee 🦋
074 Humble bee 🦋

072

073

Humble bees live in small colonies and do not produce honey, instead making "bee bread" from a mixture of nectar and pollen. This means that they have never been pursued with the same zeal as their much more prolific cousin, the honeybee. Both belong to the family Apidae, and apiculture (beekeeping) is probably the oldest branch of entomology. Archeological records tell us that pottery hives were used in Middle Eastern cultures from about 5000 BC onward. Even during the Dark Ages, a few surviving books tell us of the hives held in monasteries. However, it was not until the 15th century that keepers made serious attempts to remove their honey without killing the bees.

074

"Burly, dozing humble-bee,
Where thou art is clime for me."
RALPH WALDO EMERSON

Class	INSECTA		
Sub-Class		PTERYGOTA	
Division			ENDOPTERYGOTA
Order			

075

076

077

● 078

Aristotle believed that the queen bee, workers, and drones (males) found in bee hives were all of differing species, and that honey dropped from the air, especially "at the rise of the stars and when a rainbow descends." The English introduced the honeybee into North America in 1670 at Boston. Soon afterwards a hurricane swept the bees over the mountains, from where they colonized the rest of the continent. Frank Cowan tells us in 1865 that the Native Americans called them "English Flies" and considered them to be the "harbingers of the white man as the buffalo is of the red man…" Rather prophetically, they went on to say "…in proportion as the bee advances, the Indians and the buffalo retire."

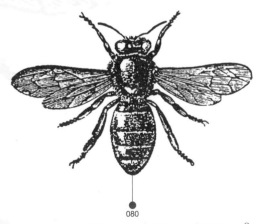

079

080

Class	INSECTA		
Sub-Class		PTERYGOTA	
Division			ENDOPTERYGOTA
Order			

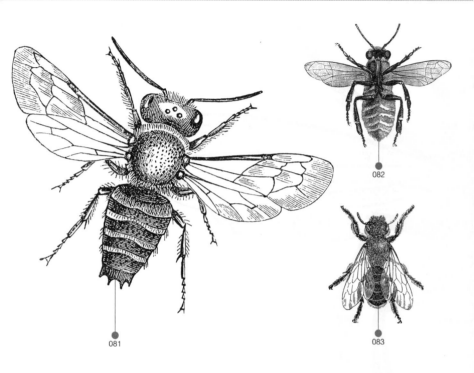

081 *Coelioxys simplex* (male)
082 Unidentified bee ⚥
083 Unidentified solitary bee ⚥
084 Unidentified solitary bee ⚥
085 *Caelioxys simplex* (female)
086 *Melecta armata* (female)
087 *Osmia rufa* (female)
088 *Megachile centuncularis* (female)

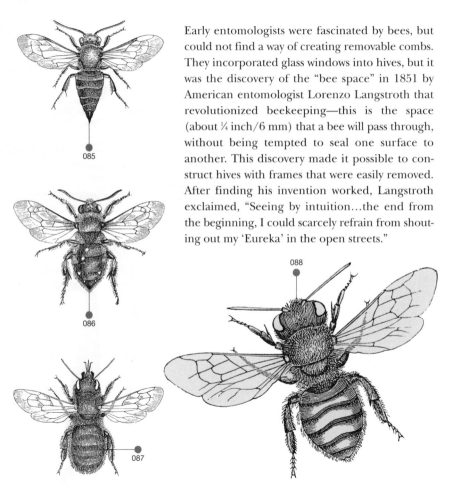

Early entomologists were fascinated by bees, but could not find a way of creating removable combs. They incorporated glass windows into hives, but it was the discovery of the "bee space" in 1851 by American entomologist Lorenzo Langstroth that revolutionized beekeeping—this is the space (about ¼ inch/6 mm) that a bee will pass through, without being tempted to seal one surface to another. This discovery made it possible to construct hives with frames that were easily removed. After finding his invention worked, Langstroth exclaimed, "Seeing by intuition…the end from the beginning, I could scarcely refrain from shouting out my 'Eureka' in the open streets."

085

086

087

088

Class	INSECTA
Sub-Class	PTERYGOTA
Division	ENDOPTERYGOTA
Order	

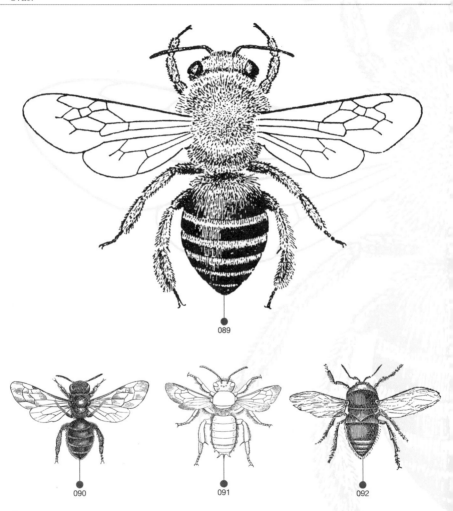

089

090

091

092

089 Solitary bee—*Andrena* sp ⚈
090 *Halictus rubicundus* (female)
091 *Anthidium manicatum* (male)
092 Carpenter bee—*Xylocopa* sp ⚈
093 *Halictus* sp ⚈ (head of male)
094 *Dasypoda* sp ⚈ (head of male)
095 *Dasypoda* sp ⚈ (leg of male)
096 *Halictus* sp ⚈ (abdomen of male)
097 *Dasypoda hirtipes* (female) ⚈

Honeybees have often been kept as pollinators as well as honey producers, but recently entomologists have discovered that they can be very choosy about the plants from which they gather pollen. The bees on this and the previous page are all solitary bees. As the name suggests, they do not live in large family groups; instead, each female will build her own nest—usually a tunnel in the ground or an existing hole in wood—and provision it with honey and pollen, then plaster it up with a mixture of mud or chewed leaves. These bees are excellent pollinators, and today entomologists are persuading farmers around the world to encourage them into fields and orchards by providing them with artificial nesting sites.

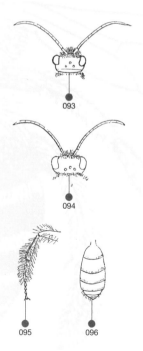

093

094

095   096

097

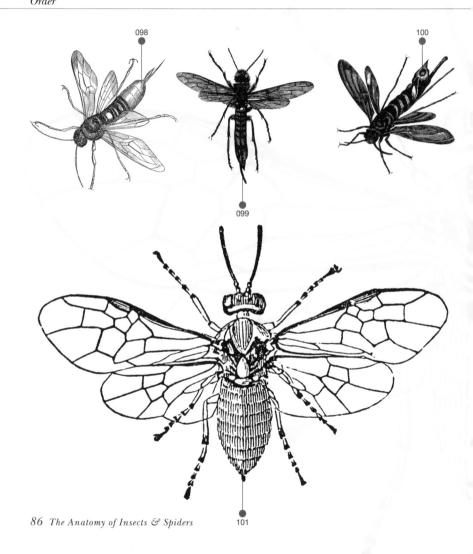

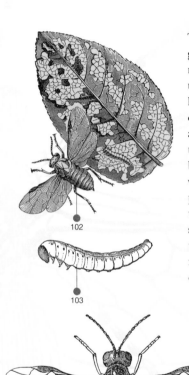

102

103

The sawflies are hymenopterans that entomologists call the Symphyta, from the Greek *symphyein*, meaning "to grow as a single stem"—a reference to the fact that this sub-order is not "wasp-waisted" like the much larger sub-order Apocrita, which comprises the rest of the order Hymenoptera and includes the sawflies' more recent cousins the bees, wasps, and ants. For some time it was believed that sawflies eat only plants; however, we now know that a few species are carnivorous or parasitic. Sawflies are the first hymenopterans to appear in the fossil record. They are called sawflies because of the egg-laying organ (ovipositor), which is not used as a stinger, but resembles a minute saw for cutting into leaves or stems in order to insert eggs.

098 Horntail or great sawfly—*Urocerus gigas*
099 Unidentified sawfly
100 Unidentified sawfly
101 Turnip sawfly—*Athalia rosae*
102 Sawfly—*Eriocampa limacina*
103 Rose sawfly larva—*Blennocampa pusilla*
104 Rose sawfly—*Blennocampa pusilla*

104

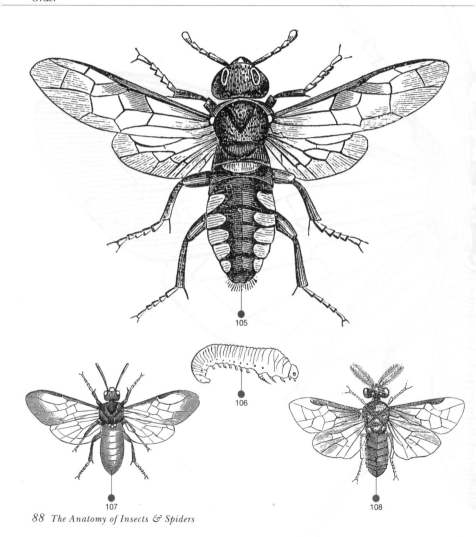

105

106

107

108

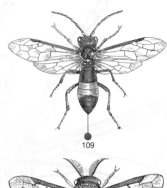

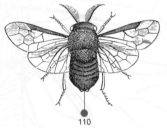

109

110

111

Most of the adult sawfly species feed on nectar or pollen and include some of the largest hymenopterans. In a few species the larvae are legless, but most of them closely resemble butterfly and moth caterpillars. They can be distinguished from the latter by counting the number of short, stumpy legs (known as prolegs) that occur behind the three pairs of longer legs on the front of the body. Butterfly and moth caterpillars never have more than five pairs of prolegs, while sawfly caterpillars never have fewer than six pairs, but may sometimes have as many as 20. Most of the caterpillars feed openly on the leaves of the food plant, but the legless species bore into stems or mine into leaf tissue.

105 *Cimbex lutea*
106 *Cimbex* sp larva ⚦
107 Sawfly—*Arge rosae* ⚦
108 *Cryptus pallipes*
109 *Tenthredo zonata* ⚦
110 *Diprion pini* ⚦
111 *Lophyrus* sp larva ⚦
112 *Lophyrus* sp cocoon ⚦
113 *Lophyrus* sp male antenna ⚦
114 *Lophyrus* sp female antenna ⚦

112

113

114

Class	INSECTA		
Sub-Class		PTERYGOTA	
Division			ENDOPTERYGOTA
Order			

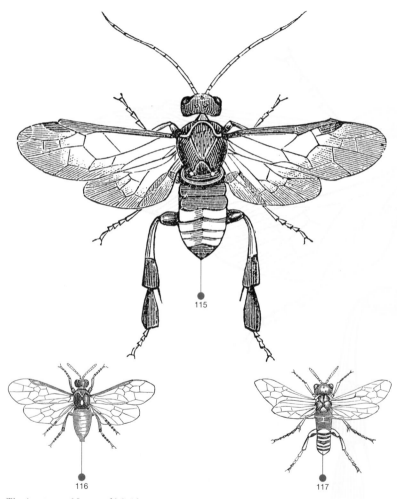

115

116

117

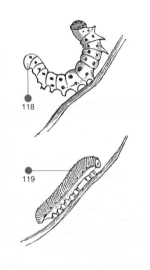

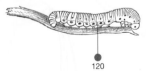

Some sawfly caterpillars are slimy to the touch, causing early entomologists to name them slugworms. A few species are serious pests in some parts of the world, including the pear slugworm, the strawberry sawfly, and the turnip sawfly. Despite this, they rarely attracted the attention of ancient entomologists, although by the 19th century the Victorians had learned that sawfly caterpillars were extremely vulnerable when they were molting. If dislodged at this stage, they would die because they could not wriggle out of their old skin. Young boys were employed to control turnip sawflies: They "would soon exterminate the caterpillars by lightly sweeping over the Turnips once or twice a week with a light broom of slender twigs."

115 *Croesus septentrionalis*
116 *Athalia spinarum*
117 *Allantus scrophularia*
118 *Croesus* sp larva
119 *Athalia* sp larva
120 *Allantus* sp larva
121 Larvae of rose sawfly—*Arge rosae*

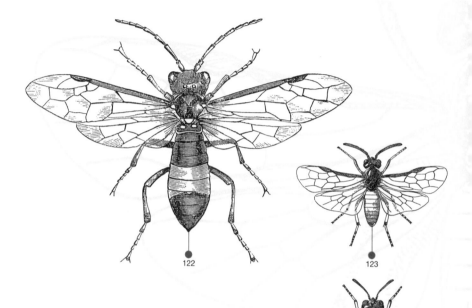

122   *Tenthredo zonata*

123   Sawfly (male)—*Eriocampa rosae*

124   Sawfly (female)—*Eriocampa rosae*

125   Rose slug-worm larva—*Eriocampa rosae*

126   Rose sawfly—*Emphytus cinctus*

127   Rose sawfly larva—*Emphytus cinctus*

128   Gooseberry and currant sawfly—*Nemetus ribesii*

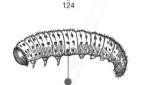

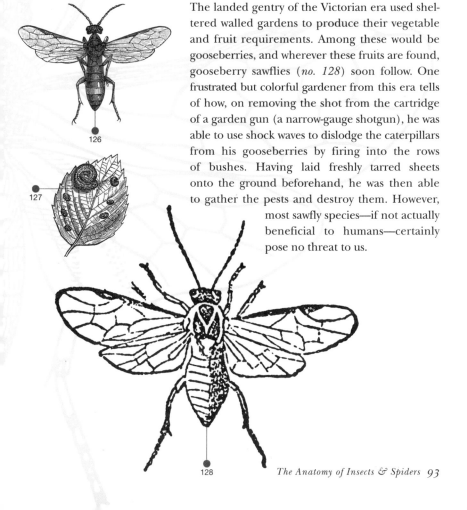

126

127

The landed gentry of the Victorian era used sheltered walled gardens to produce their vegetable and fruit requirements. Among these would be gooseberries, and wherever these fruits are found, gooseberry sawflies (*no. 128*) soon follow. One frustrated but colorful gardener from this era tells of how, on removing the shot from the cartridge of a garden gun (a narrow-gauge shotgun), he was able to use shock waves to dislodge the caterpillars from his gooseberries by firing into the rows of bushes. Having laid freshly tarred sheets onto the ground beforehand, he was then able to gather the pests and destroy them. However, most sawfly species—if not actually beneficial to humans—certainly pose no threat to us.

128

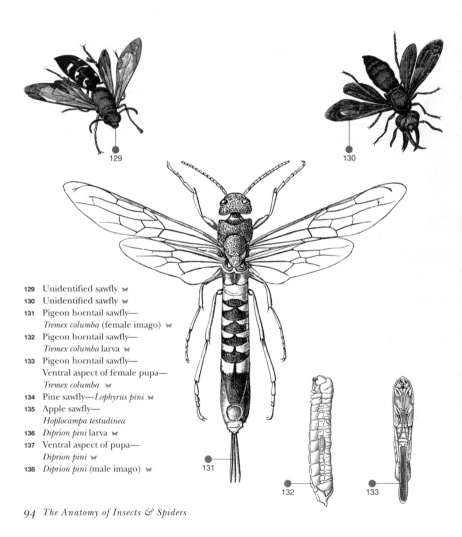

129 Unidentified sawfly 🪶

130 Unidentified sawfly 🪶

131 Pigeon horntail sawfly—
Tremex columba (female imago) 🪶

132 Pigeon horntail sawfly—
Tremex columba larva 🪶

133 Pigeon horntail sawfly—
Ventral aspect of female pupa—
Tremex columba 🪶

134 Pine sawfly—*Lophyrus pini* 🪶

135 Apple sawfly—
Hoplocampa testudinea

136 *Diprion pini* larva 🪶

137 Ventral aspect of pupa—
Diprion pini 🪶

138 *Diprion pini* (male imago) 🪶

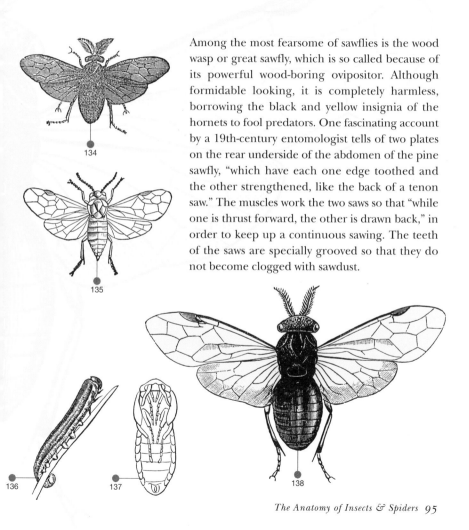

Among the most fearsome of sawflies is the wood wasp or great sawfly, which is so called because of its powerful wood-boring ovipositor. Although formidable looking, it is completely harmless, borrowing the black and yellow insignia of the hornets to fool predators. One fascinating account by a 19th-century entomologist tells of two plates on the rear underside of the abdomen of the pine sawfly, "which have each one edge toothed and the other strengthened, like the back of a tenon saw." The muscles work the two saws so that "while one is thrust forward, the other is drawn back," in order to keep up a continuous sawing. The teeth of the saws are specially grooved so that they do not become clogged with sawdust.

134

135

136

137

138

THRIPS, TERMITES, DRAGONFLIES, & MAYFLIES

Thysanoptera, Isoptera, Odonata, & Ephemeroptera

Thrips are in the relatively small order of Thysanoptera, which consists of around 5,000 known species. The Greek word *thysanos* means "a fringe" and refers to their slim wings, which are lined with hairs. Despite their diminutive size, thrips can be serious crop pests—a problem magnified by their worldwide distribution.

Termites, in the highly social order of Isoptera, are often referred to as white ants. This description is slightly off the scientific mark, given that ants are in a different insect order, although it originates from the anemic appearance of the workers. The Isoptera cannot be mentioned without reference to the caste system, which comprises soldiers, many workers, and reproductive individuals. Mouthpart (mandibular) shapes of the soldiers vary greatly, giving the insect an unbalanced appearance in which the head is greatly enlarged. Termites nest in fascinating terrestial mounds or inside wood, but the former is perhaps most familiar

In Greek *odonto* means "of the teeth," and in the order Odonata this refers to the fearsome teeth commonly found on the mandibles of the predatory adults. There are about 5,000–6,000 known species of Odonata, most of which inhabit the tropics. The genus name of *Libellula* was assigned to dragonflies by the great Swedish zoologist and botanist Carolus Linnaeus. And it is said that Dr. Thomas Mouffet divided them into three categories—*maxima, media,* and *minima*—in the 17th-century Latin publication *Insectorum Theatrum.* Then, in the latter part of the 18th century, Johann Fabricius added two more genera to Linnaeus's: *Aeshna* and *Agrion.*

The final order that is alluded to in this chapter is the Ephemeroptera, or mayflies.

001 Pea thrips—*Kakothrips pisivorus* ✎
002 *Phlaeothrips coriacea*
003 Thrips—*Limothrips* sp ✎
004 Thrips—*Limothrips* sp ✎
005 Pea thrips—*Kakothrips* sp ✎
006 Pea thrips larva ✎

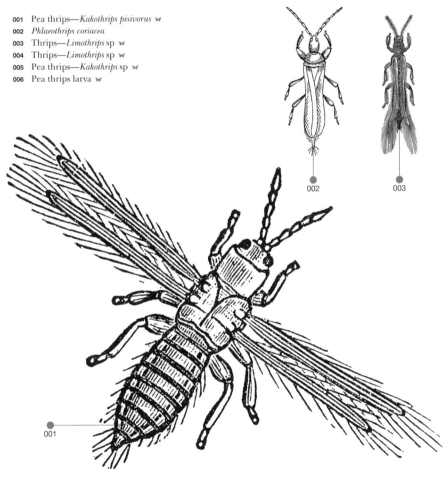

002

003

001

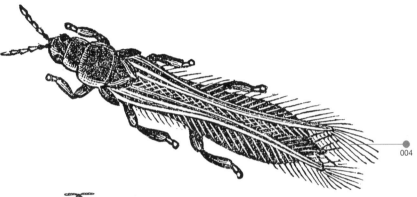

004

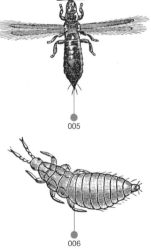

005

006

Given thrips' insignificant size, the development of the microscope in the 17th century was no doubt a great advantage in studying these insects. Thrips, never referred to in the singular, may be winged (alate) or wingless (apterous)—the flying thrips are commonly called thunder bugs or thunder flies, since they are often associated with this type of weather. Mouth parts of the Thysanoptera allow them to suck up their food. Apart from feeding on plant juices, thrips may be found on decaying organic matter or fungi; others are predators and prey upon mites, aphids, and even other thrips. Simply depicted here are the fringes of the hairs on both the hind- and forewings, which gave rise to the Greek naming of the order.

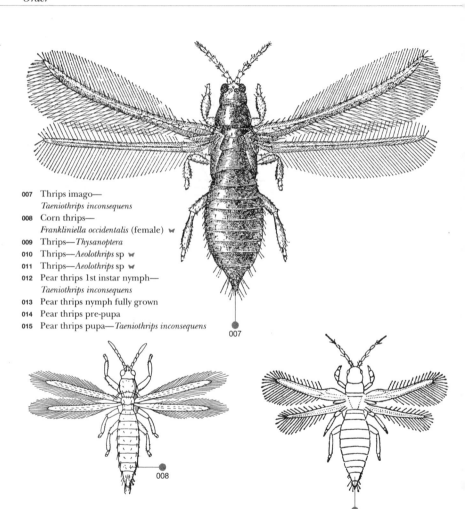

007 Thrips imago—
   Taeniothrips inconsequens
008 Corn thrips—
   Frankliniella occidentalis (female) ⋈
009 Thrips—*Thysanoptera*
010 Thrips—*Aeolothrips* sp ⋈
011 Thrips—*Aeolothrips* sp ⋈
012 Pear thrips 1st instar nymph—
   Taeniothrips inconsequens
013 Pear thrips nymph fully grown
014 Pear thrips pre-pupa
015 Pear thrips pupa—*Taeniothrips inconsequens*

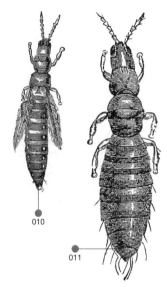

010

011

Apart from the direct damage that thrips cause though their feeding habits, certain species transmit plant viruses, and others (mostly tropical species) are associated with galls (*see Glossary, page 277*). Interestingly we are told by Margaret M. Fagan in *The Uses of Insect Galls*, published in the *American Naturalist* in 1918, that galls were considered to be supernatural phenomena during the Middle Ages, with the ability to see into the future! The nature of misfortune alluded to depended on the type of insect contained within the gall. Some Thysanoptera species tend their eggs and young, while certain thrips from Panama present a good example of a highly developed sociality. Members of this genus live communally, care for their broods, and forage for their young.

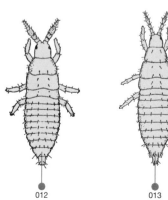

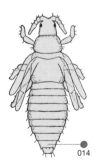

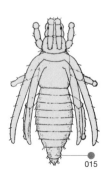

012

013

014

015

Class	INSECTA		
Sub-Class		PTERYGOTA	
Division			EXOPTERYGOTA
Order			

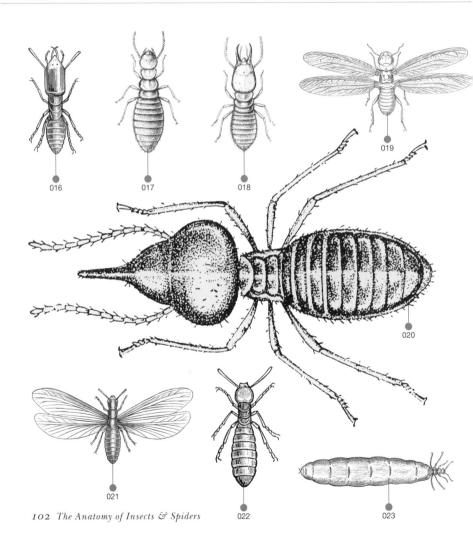

016 Soldier termite
017 Worker termite—
   Prorhinotermes simplex
018 Soldier termite—
   Prorhinotermes simplex
019 Winged adult termite—
   Hodotermes mossambicus
020 Soldier termite—*Nasutitermes* sp
021 Queen before shedding wings
022 Worker termite
023 Queen after shedding wings,
   abdomen distended with eggs
024 Worker termite—
   Mastotermes darwiniensis (male)
025 Winged adult termite—
   Termes lucifugus 🦋
026 Older substitution queen—
   Termes lucifugus
027 Winged adult termite—
   Kalotermes flavicollis
028 Eyed, grass cutting termite soldier—
   Hodotermes havilandi
029 Eyed, grass cutting termite worker—
   Hodotermes havilandi

The highly organized and social termites, or white ants, live mostly in hot climes such as Africa and South America, although some are found in the warmer areas of Europe. Of particular fascination is the regulation of temperature and carbon dioxide enabled by the intricate internal design of the mounds belonging to fungus-farming termites. And the "magnetic mounds" of certain North Australian species are cleverly constructed so that the broad surface of the mound is exposed to the early and late sun, with the narrow aspect exposed to the hot midday sun. The distended body of the fertilized, apterous queen termite (*see no. 023*), can reach up to a massive 3½ inches/9 cm—making an attractive food source in some areas of Africa.

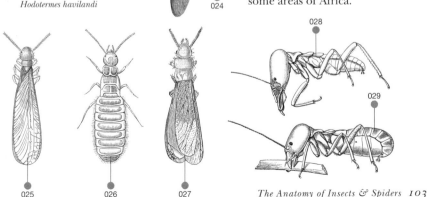

Class	INSECTA
Sub-Class	PTERYGOTA
Division	EXOPTERYGOTA
Order	

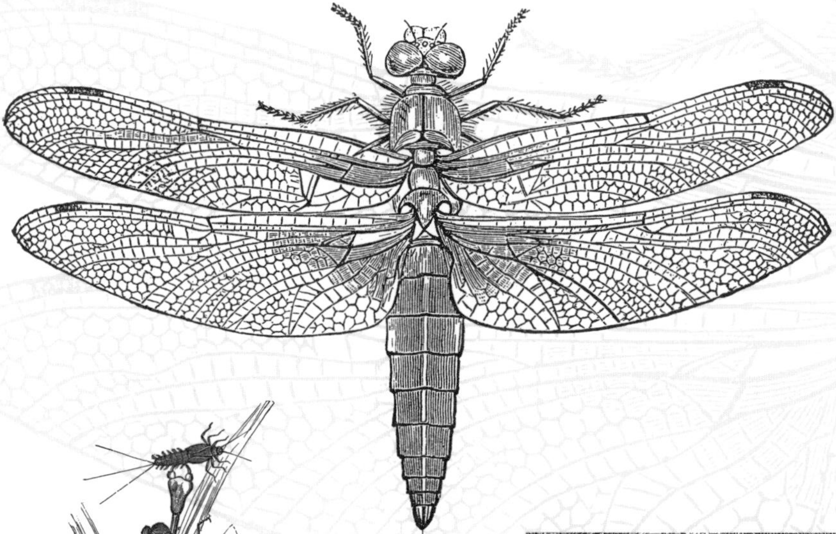
031

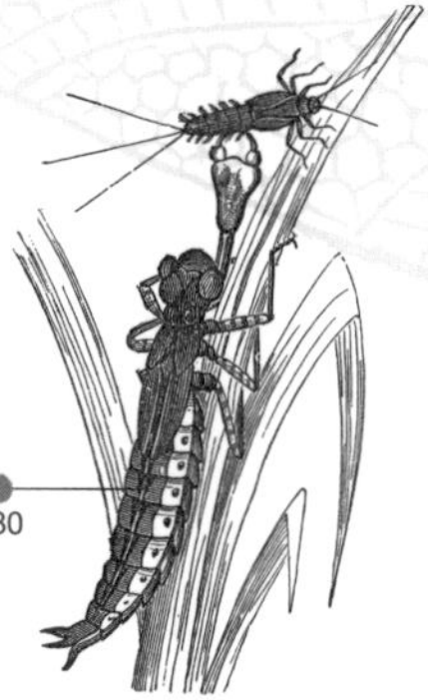
030

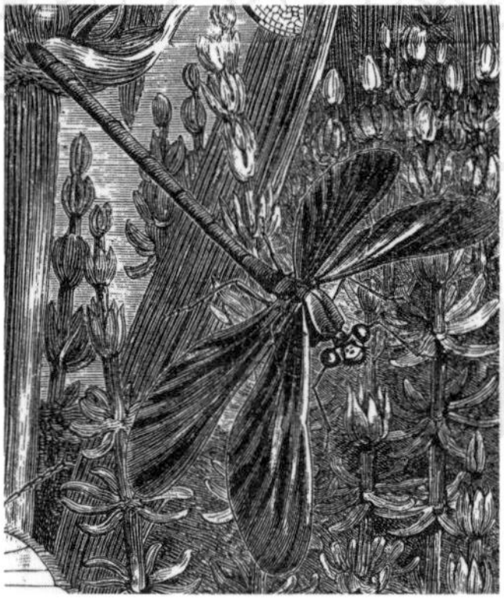
032

030 Dragonfly grub
031 Unidentified dragonfly
032 Beautiful demoiselle darter—*Calopteryx virgo*
033 *Libellula depressa*
034 *Libellula* sp emerging from pupa case
035 *Libellula* sp larva

 033

Confusingly, it is common for the English to use the term "dragonfly" in reference both to the order Odonata and the sub-order Anisoptera (hence, in the following pages, this term refers to damselflies as well as dragonflies). The Greek *anisoptera* means "unequal wings," and the front wings of these Odonata are narrower than the hindwings. To capture its prey, the dragonfly uses the tremendous structure of its "mask" (claws on an extending limb). When folded, this covers most of the underside of its head, hence the term. Some dragonflies are known as darters, because they dart toward their prey; and we are told by T. D. A. Cockerell, in a paper entitled "Dru Drury, An Eighteenth Century Entomologist" printed in *Scientific Monthly* in 1922, that the said entomologist referred to "Libella" nymphs as cats.

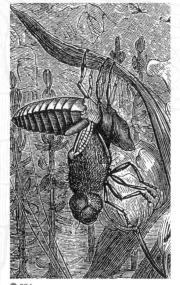

034

035

Class	INSECTA
Sub-Class	PTERYGOTA
Division	EXOPTERYGOTA
Order	

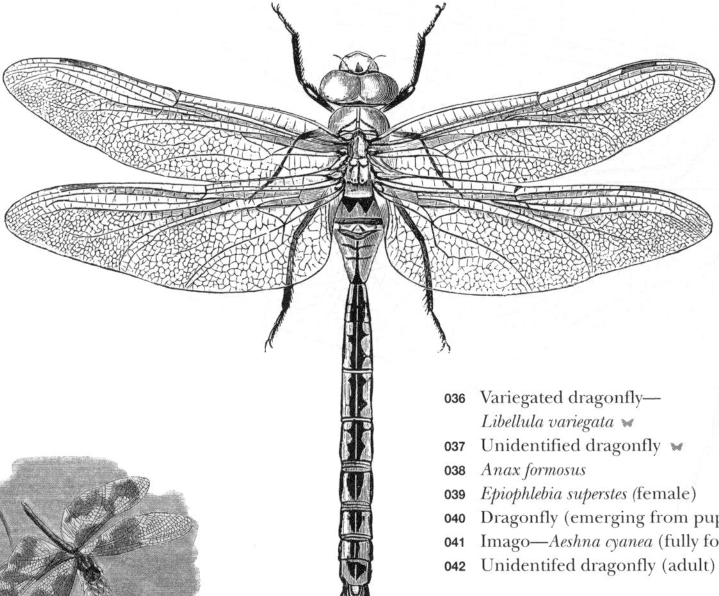

038

036 Variegated dragonfly—*Libellula variegata*
037 Unidentified dragonfly
038 *Anax formosus*
039 *Epiophlebia superstes* (female)
040 Dragonfly (emerging from pupa case)
041 Imago—*Aeshna cyanea* (fully formed)
042 Unidentifed dragonfly (adult)

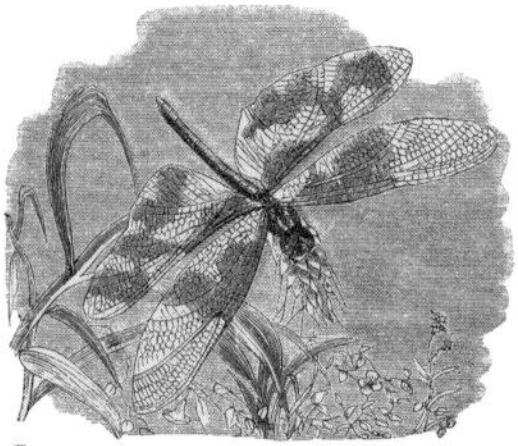

036

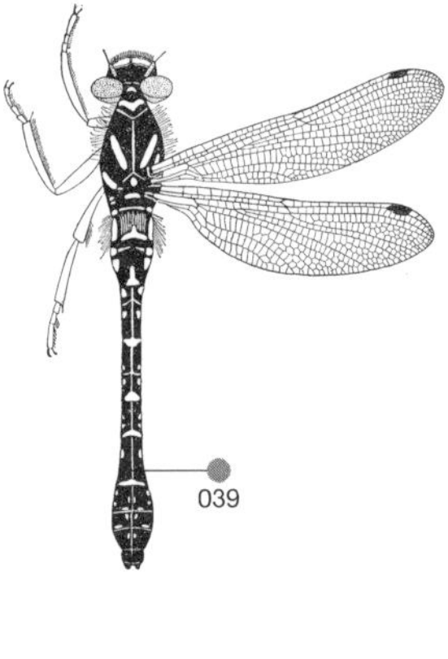

039

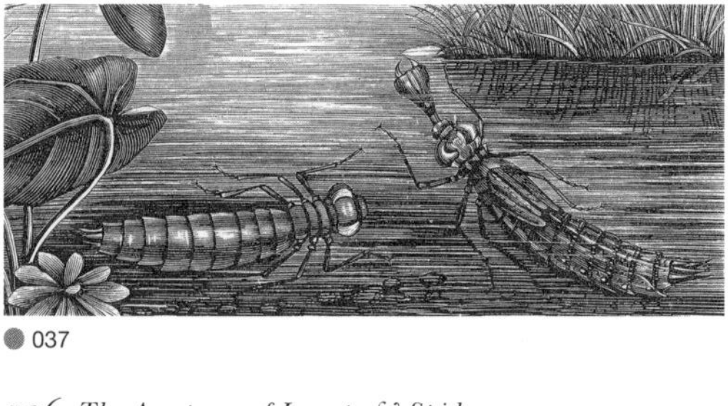

037

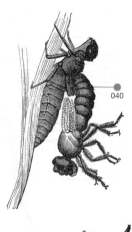

040

The larvae of true dragonflies may have a colossal 15 aquatic juvenile stages, or instars, and in his 18th-century publication *Exposition of English Insects,* Moses Harris is said to have mistakenly identified some juveniles as different species—a mistake he also repeated for some male and female Odonata. Nonetheless Harris, along with other writers of his time, promoted interest in the study of these beautiful day flyers. Mating in this group is a curious affair: Not only do the creatures adopt a tandem position, but in 1979 it was discovered by Jonathan K. Waage, and reported in *Science* under "Dual Function of the Damselfly Penis: Sperm Removal and Transfer," that the male can remove the sperm of previous mates of his female partner.

041

042

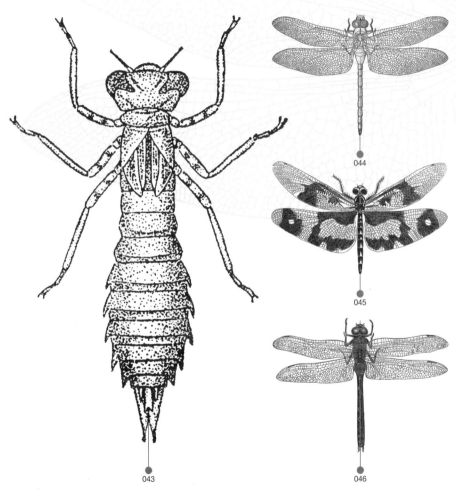

043

044

045

046

047

"...these insects are, indeed, the dragons of the air, far more voracious and active than even the fabled dragons of antiquity."
REV. J. G. WOOD

These ancient insects have beautifully veined wings that move as independent pairs, unlike the tendency toward connected movement that insect evolution has seen. Fossil records indicate that wingspans for ancestors of these creatures were in the region of an immense 27–31 inches/ 70–80 cm. Unjustly named "horse-stingers," dragonflies seen darting toward horses were thought to sting them. These slender insects were also named "devil's darning needles" in Europe, but were more favorably thought of as the spirits of rice plants in Japan, or as tokens of good luck. And we are told that in southwest America dragonflies were painted onto pottery to symbolize life. Prominent features of the dragonflies include their huge eyes, which are unsurpassed by those of any other insect orders, and their well-developed flight muscles.

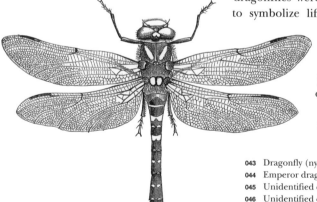

043 Dragonfly (nymph)
044 Emperor dragonfly—*Anax imperator* 🦋
045 Unidentified dragonfly 🦋
046 Unidentified dragonfly 🦋
047 Unidentified dragonfly 🦋
048 Hawker dragonfly—*Cordulegaster annulatus* 🦋

048

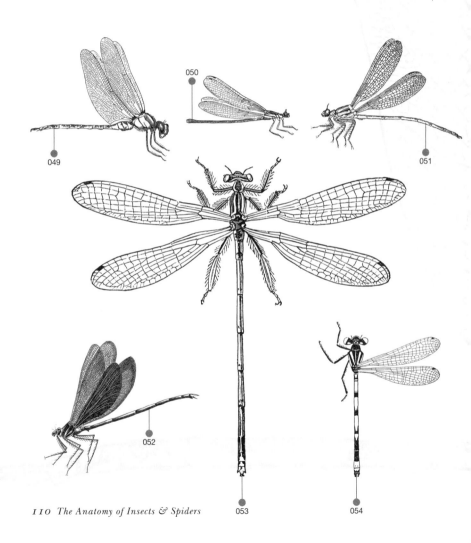

Some insects pass through a pupal or chrysalid stage in their life histories, and others do not. The latter group is often referred to as the older one, and it is to this group that the dragonflies and damselflies belong. Damselflies are part of the sub-order Zygoptera, a Greek word meaning "paired wing," referring to the linked movements in the two sets. The femininely named and delicately formed damselflies dance agitatedly in the air rather than fly. They are often seen to fly together as male and female pairs when the female is laying her eggs. The three slender gills (caudal lamellae) that project from the abdomen of the damselfly nymph make this sub-order easily distinguishable from dragonfly nymphs.

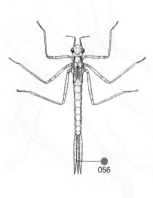

056

055

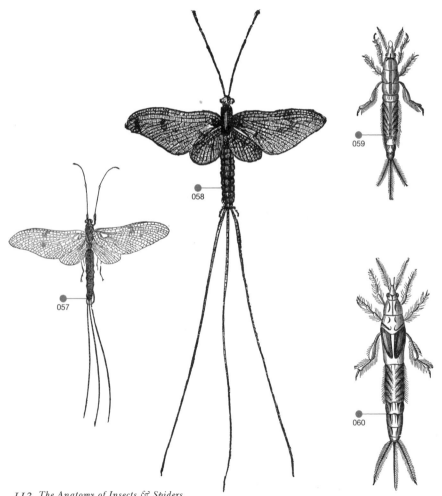

MAYFLIES *1*

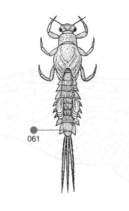

061

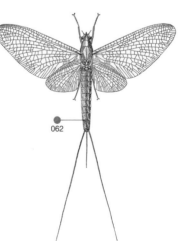

062

Mayflies belong to the ancient group of winged insects known as the Palaeoptera, which includes both dragonflies and damselflies. They are misleadingly named, for these insects are also seen outside the month of May. The models used by anglers are in the order Ephemeroptera (mayflies)—the Greek word *ephemeros* means short-lived and refers to the adult "spinners," which may live for less than a day or up to a week. Incidentally, this life stage is formed after the molt of a winged individual—a unique occurrence in the insect world. Benjamin Franklin, the 18th-century American publisher, printer, statesman, and scientist, referred to these insects in an essay that he wrote to Madame Brillon in 1778, entitled *Soliloquy of a Venerable Ephemera Who Had Lived Four Hundred and Twenty Minutes, or The Ephemera.*

057 Unidentified mayfly ✎
058 Mayfly—*Ephemera danica* ✎
059 Mayfly larva
060 Mayfly pupa
061 Mayfly nymph
062 Unidentified mayfly (adult) ✎

BUTTERFLIES *Lepidoptera*

Butterflies form part of the colossal order of Lepidoptera, a captivating collection of approximately 150,000-plus known species. The Greek word *lepis* means "scale"; thus Lepidoptera means "scale wings." Members of this popular group live in diverse habitats, ranging from tropical rain forests to cooler mountain ranges. While there was a flurry of activity in the 18th century among collectors of beautiful butterfly specimens in the West, we are told that natives of the tropics had a very different attitude toward them and regarded some butterflies as their ancestors' souls. The presumed demand, and the invention of the butterfly net (which was apparently in use by the early 1700s), indicates a widespread interest in the collection of butterflies. Early Lepidopteran traps included the cumbersome clap net, bat-fowling net, or batfolder, which was beautifully depicted in *The Aurelian* by Moses Harris in 1766, and later reproduced in

Dr E. B. Ford's *Butterflies* in 1971. And in *Butterfly Collectors and Painters* (2000) Pamela Gilbert relates the practice of "lepidochromy," whereby actual wings were used to create illustrations.

The English society devoted to the study of these outstanding insects employed the name Aurelian in its title. The name originates from a particular butterfly chrysalis, which is golden and hence named *aureolia*. By the 1740s the Society of Aurelians was said to be thriving. The striking colors and incredible patterns on butterfly wings no doubt seduced people into becoming collectors. Certainly an artist's palette would want for nothing if clothed with all the beautiful colors of butterflies—hues created by their complex, delicate scales and the way in which the light dances across them. Indeed, the butter-colored wings of the brimstone butterfly are thought to have inspired the common name for this part of the order.

Class	INSECTA
Sub-Class	PTERYGOTA
Division	ENDOPTERYGOTA
Order	

001

002

003

004

Lepidoptera BUTTERFLIES 1

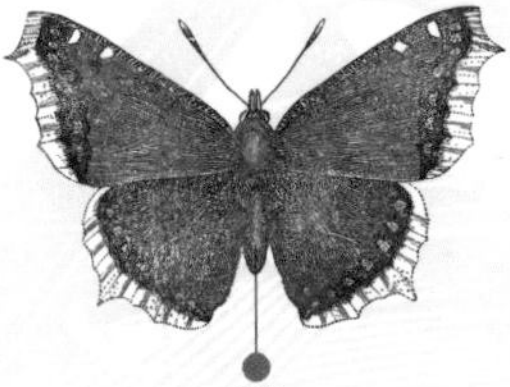

005

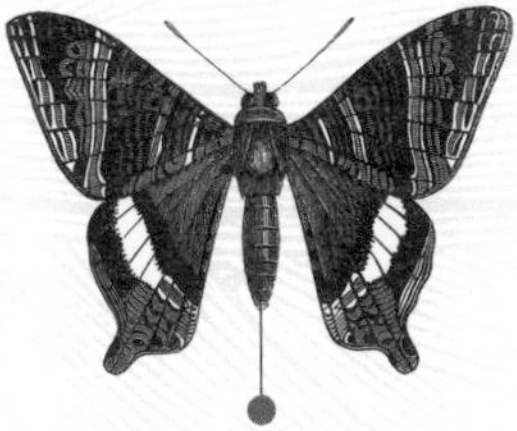

006

Among other families of butterflies, Henry Bates mentioned the family Acraeinae in conjunction with the phenomenon of mimicry, based on his observations in the tropics of South America. Butterflies of this family have just two pairs of functioning legs. The common name for the brush-footed butterfly is derived from its forelegs, which look like small brushes due to the covering of hairs. The mourning cloak butterfly is a rare immigrant in Britain, known as the Camberwell beauty (*no. 005*). London's metropolis has now engulfed Camberwell village, but as a hamlet, E. B. Ford tell us it witnessed a few sightings of this butterfly. One such citing was recorded in 1748 and published in *The Aurelian* (in 1766).

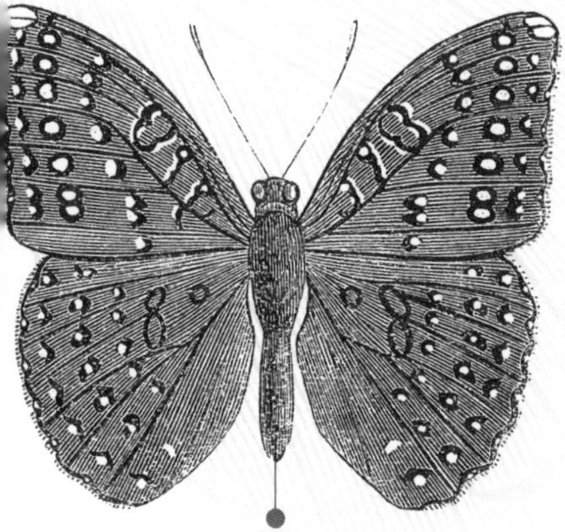

007

001 *Acraea vesta*
002 *Acraea rakeli*
003 *Acraea perenna*
004 *Aterica rabena*
005 Camberwell beauty—*Nymphalis antiopa*
006 *Acraea egina*
007 *Aterica meleagris*
008 *Myrina jafra*

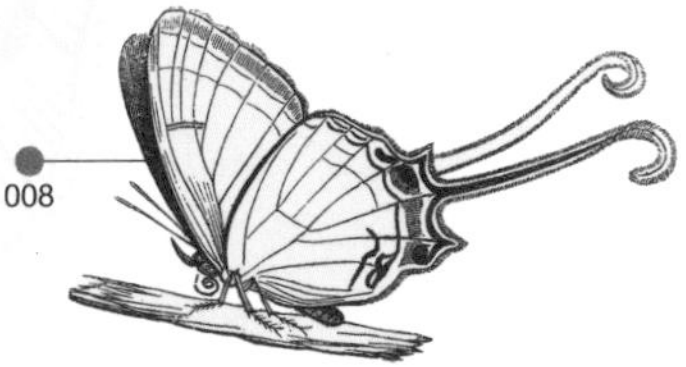

008

009

010

011

012

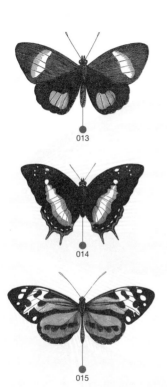

013

014

015

Entomological history is peppered with stories relating to fraudulent claims where insects of great value were said to be held in the narrator's possession. The seductive Camberwell beauty is no exception (*see page 117*): Dr. Ford recounts a story where the yellow borders of the wings were covered in white paint, because it was an old and misguided belief that the latter color indicated a rare native British specimen. It is said that the transformed beauty ended up in the British Museum. The genus *Limenitis* is assigned to the admiral butterflies, although Moses Harris wrote "Admirable" as opposed to "Admiral" in *The Aurelian.* This grand tome, with a beautifully handwritten index in the first edition, includes exquisite color plates—a must-see for all.

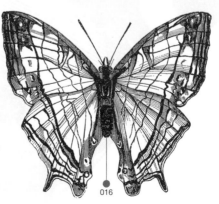

016

Class	INSECTA	
Sub-Class		PTERYGOTA
Division		ENDOPTERYGOTA
Order		

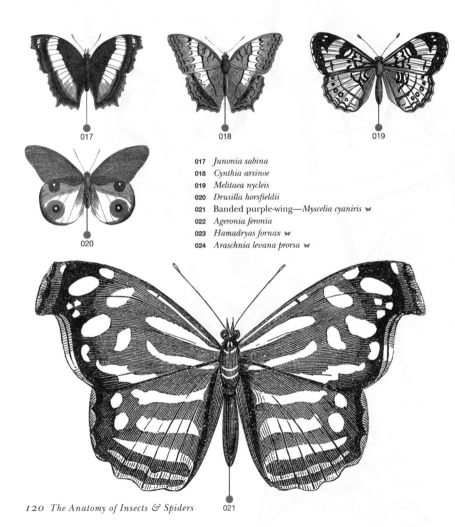

017 *Junonia sabina*
018 *Cynthia arsinoe*
019 *Melitaea nycleis*
020 *Drusilla horsfieldii*
021 Banded purple-wing—*Myscelia cyaniris* ✎
022 *Ageronia feronia*
023 *Hamadryas fornax* ✎
024 *Araschnia levana prorsa* ✎

022

023

The beautifully named genus of *Cynthia* contains the butterfly ladies: the painted lady, the American painted lady, and the West Coast lady; the painted lady has been named *Belle Dame* by the French. The pioneering Charles Darwin, who embarked on his famous voyage aboard the *Beagle* in 1831, was in Brazil when he first heard—and subsequently enlightened his peers to—the noise made by a certain butterfly, *Ageronia feronia* (*no. 022*). He chronicled this sound as being similar to "a toothed wheel passing under a spring catch." It was actually the male "crackers" or "clicking" butterflies of this species that produced the sound.

024

Class	INSECTA	
Sub-Class		PTERYGOTA
Division		ENDOPTERYGOTA
Order		

025

026

027

028

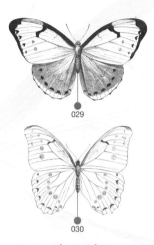

029

030

The magnificent tropical Morphidae are exceptionally efficient flyers—unsurprisingly, given their large wing size. The luster and coloring of these structures not only made them highly attractive to collectors in an intact state, but also prompted masks and ornaments to be fashioned from them, and trade among chic Europeans maintained this art as a thriving business. As well as the final product being of great interest to us, the complete metamorphosis of the Lepidopteran life cycle was observed with a fascination which echoed that of the ancient Egyptians. In his 1865 publication, Frank Cowan chronicled that this society drew an analogy between the human soul and the release of the winged insect from its chrysalis.

031

025 Malay lacewing—*Cethosia hypsea* ✷
026 *Libythea fulgurata* ✷
027 Large yeoman—*Cirrochroa aoris* ✷
028 *Colaenis dido* ✷
029 Mother of pearl morpho—*Morpho laertes* ✷
030 White morpho—*Morpho polyphemus* ✷
031 Pasha—*Herona marathus* ✷
032 *Caerois chorineus*

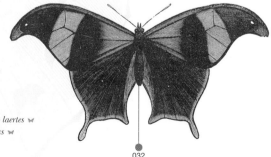

032

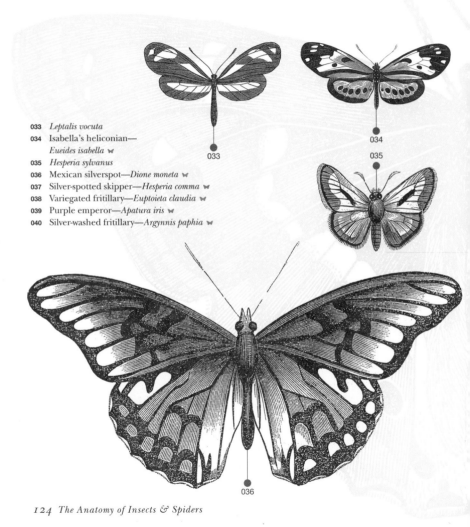

033 *Leptalis vocuta*
034 Isabella's heliconian—
     Eueides isabella 🦋
035 *Hesperia sylvanus*
036 Mexican silverspot—*Dione moneta* 🦋
037 Silver-spotted skipper—*Hesperia comma* 🦋
038 Variegated fritillary—*Euptoieta claudia* 🦋
039 Purple emperor—*Apatura iris* 🦋
040 Silver-washed fritillary—*Argynnis paphia* 🦋

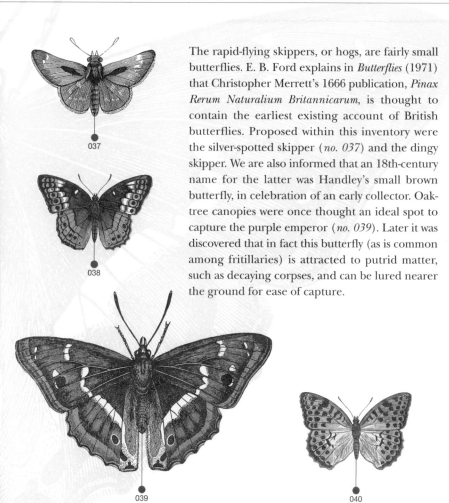

The rapid-flying skippers, or hogs, are fairly small butterflies. E. B. Ford explains in *Butterflies* (1971) that Christopher Merrett's 1666 publication, *Pinax Rerum Naturalium Britannicarum*, is thought to contain the earliest existing account of British butterflies. Proposed within this inventory were the silver-spotted skipper (*no. 037*) and the dingy skipper. We are also informed that an 18th-century name for the latter was Handley's small brown butterfly, in celebration of an early collector. Oak-tree canopies were once thought an ideal spot to capture the purple emperor (*no. 039*). Later it was discovered that in fact this butterfly (as is common among fritillaries) is attracted to putrid matter, such as decaying corpses, and can be lured nearer the ground for ease of capture.

037

038

039

040

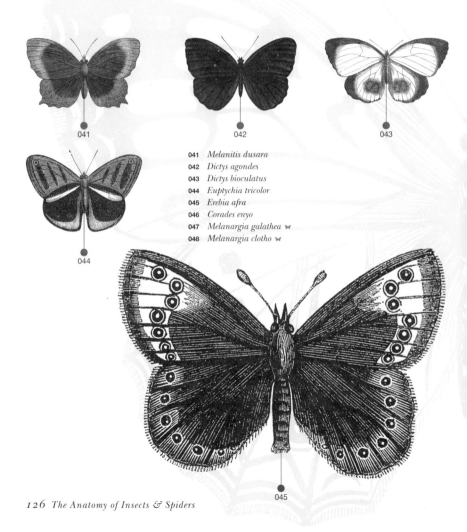

041   *Melanitis dusara*
042   *Dictys agondes*
043   *Dictys bioculatus*
044   *Euptychia tricolor*
045   *Erebia afra*
046   *Corades enyo*
047   *Melanargia galathea* ✎
048   *Melanargia clotho* ✎

041

042

043

044

045

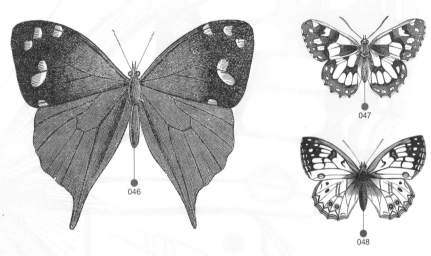

046

047

048

*"I look for butterflies
That sleep among the wheat:
I make them into mutton-pies,
And sell them in the street."*
CHARLES LUTWIDGE DODGSON
(A.K.A. LEWIS CARROLL)

The marbled white butterfly is a member of the Satyridae family, which has earned the collective name of "the browns"; however, brown is widespread, but by no means the only shade apparent in this group. While dragonflies represent an ancient group of insects, butterflies and moths are thought to have appeared much later in the Tertiary period, which is thought to have witnessed the rise of mammals and the decline of the dinosaurs. The marbled white was another butterfly included in Christopher Merrett's 1666 list, but E. B. Ford indicates that this species probably arrived in Britain in the later Holocene period.

Class INSECTA

Sub-Class PTERYGOTA

Division ENDOPTERYGOTA

Order

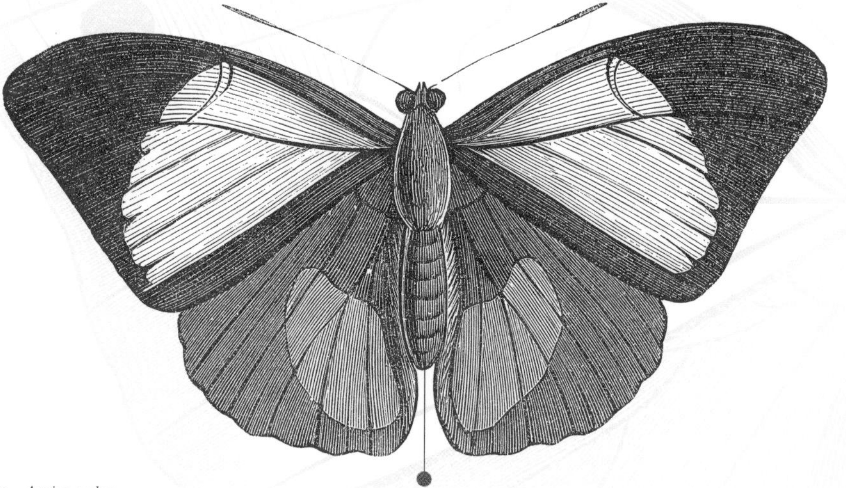

049

049 *Agrias aedon*
050 *Eryphanis reevesii*
051 Orange owl butterfly—*Opsiphanes boisduvalii*
052 *Dasyophthalma rusina*
053 *Protogonius cecrops*
054 Cocoa mort blue—*Caligo teucer*
055 *Morpho* sp
056 *Hypna clytemnestra*

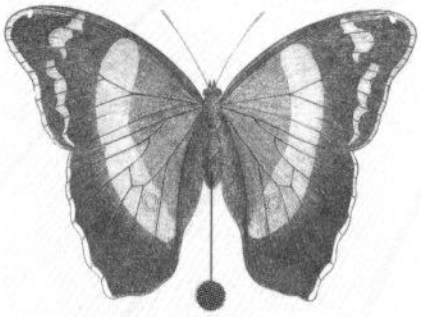

050

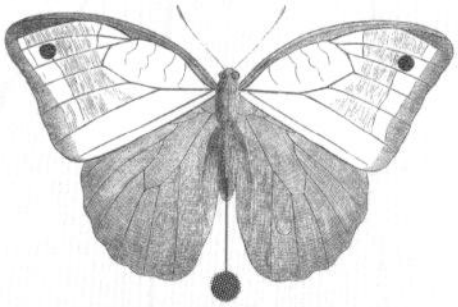

051

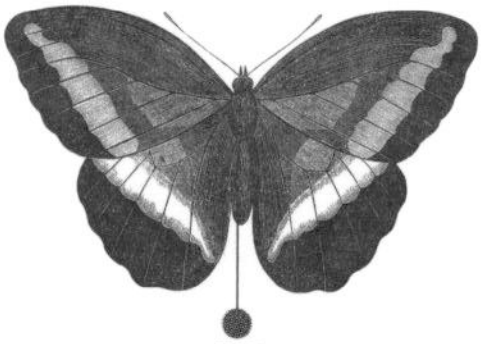

052

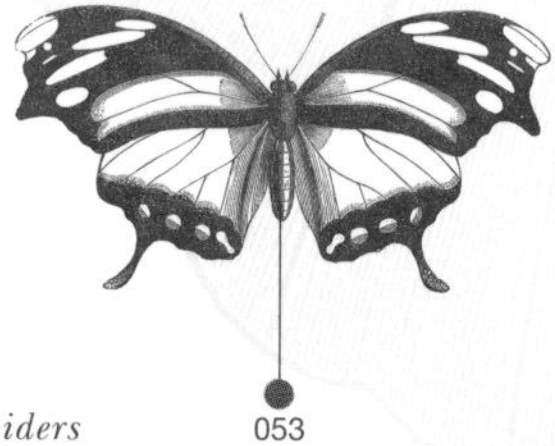

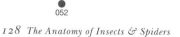

053

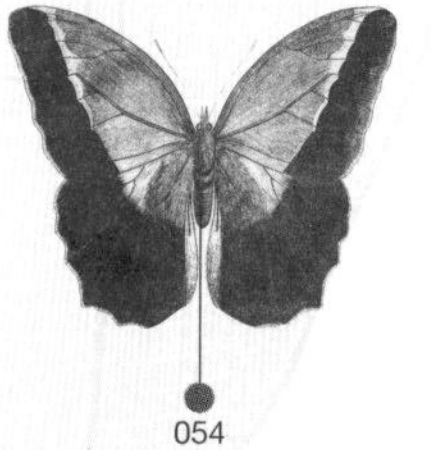

054

The neo-tropical Brassolinae are at home in the South American jungles. *Caligo* species of this group are commonly known as owl butterflies—the stunning "eyes" on the underside of the large wings being responsible for this common name. These *faux* features serve as a deterrent against predators, which are startled by their sudden exposure. Another way in which particular butterflies might protect themselves against an enemy is by matching coloration and patterning with their habitat. In this way, if they are at rest on a branch, for example, then they will be camouflaged and hopefully evade attempted consumption.

055

"The toad beneath the harrow knows
Exactly where each tooth-point goes;
The butterfly upon the road
Preaches contentment to that toad."
Rudyard Kipling

056

057   Great nawab—*Polyura eudamippus* ✎

058   Jewelled nawab—*Polyura delphis* ✎

059   *Idea agelia*

060   *Lycorea atergatis*

061   *Ituna phenarete*

062   *Colaenis pherusa*

063   *Colaenis euchroia*

064   *Erycina licarsis*

The history of insect images, including the original woodcut forms, is undeniably fascinating, some historical illustrations have such detail that little is left to the imagination. For instance, the colorful illustrations of Lepidoptera in credited works (such as that of Moses Harris in 1766) fall into this category. Conversely, images such as those in *The Theater of Insects* in 1658 are very crude, if endearing, given the asymmetrical wings of some Lepidoptera in both shape and pattern; they are accompanied by antiquarian text such as "the Butterfly hath a two forked beak or bill, and within those forks is couched another little bill or beak, with which they suck in, some the day dew, others the night."

061

062

063

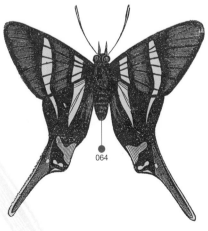

064

065

066

067

068

069

070

071

065   Black-veined white—*Aporia crataegi* ﹅
066   Orange tip—*Anthocharis cardamines* ﹅
067   *Anthocharis eupheno*
068   *Helicopis gnidus*
069   Vine-leaf vagrant—*Eronia cleodora* ﹅
070   Pale clouded yellow—*Colias hyale* ﹅
071   *Papilio orion*

The clouded yellow butterflies (*no. 070*) and the Palaearctic orange tips (*no. 066*) are members of the Pieridae family. A. Heath, in an article entitled "*Colias Edusa* and *Colias Hyale,*" printed in *Science* in 1892, reports a bountiful year in Britain in respect of these butterflies, given "its [*C. hyale*'s] comparative rarity here" and reminisces on a comparable abundance in 1886. A perhaps more familiar member of this family is the cabbage white, renowned for the destructive tendencies of its juvenile caterpillar stage. It is assumed to have a European origin and, in J. Eliot Moss's 1933 article "The Natural Control of the Cabbage Caterpillars *Pieris* spp.," the author describes two particular insect parasites as being useful control agents.

072

073

074

075

076

077

Again we are reminded, by a quote from *The Theater of Insects or lesser living Creatures, or Insectorum Theatrum* by Thomas Mouffet that entomological "medical" anecdotes have appeared throughout history: "The catterpillers that are upon spurges (in the opinion of Hippocrates) are very good for purulent wombs, especially if they be dried in the sun, with the double weight of dunghil Worms, and adding a little Anniseed, bringing them into powder, and infusing them in the best white wine...." This, despite the fact that one page earlier we are told (of caterpillars) that "they are all venomous"! Mouffet is listed under "the butterfly collectors" in Michael A. Salmon's *The Aurelian Legacy*, a place he shares with recognizable greats such as the English naturalist Dru Drury and two Rothschild brothers.

078

072 *Nymphidium arminius*
073 *Hypanartia kefersteini*
074 *Papilio phaedrus*
075 *Epicalia pierretii*
076 *Heliconius ricini*
077 *Diorina laonome*
078 *Diadema salmacis*

Class	INSECTA	
Sub-Class		PTERYGOTA
Division		ENDOPTERYGOTA
Order		

"Our North…countrymen call it (the moth) Saule, i.e. …the soul; because some silly people in old time did fancy that the souls of the dead did fly about in the night, seeking light."

THOMAS MOUFFET

079 *Eurycus cressida*
080 African giant swallowtail—*Papilio antimachus* ⋈
081 Tiger brown—*Orinoma damaris* ⋈
082 *Papilio coon*
083 Chalkhill blue—*Lysandra corydon* ⋈
084 Spotted Adonis blue—*Lycaena adonis* ⋈
085 *Papilio ulysse*
086 Eastern tiger swallowtail—*Papilio glaucus* ⋈

079

080

081

082

083

The huge *Papilio antimachus* (*no. 080*) decorates the tropics of Africa and, with an immense wingspan of about 10 inches/25 cm, it is—not surprisingly—the largest of the African butterflies. The nearctic tiger swallowtail, a common inhabitant of North America, has a wingspan of about 3¾ inches/9.5 cm. Some Papilionids have elegant projections from their hindwings, sometimes referred to as "spurs." In Britain swallowtails declined around the time of the First World War, when alternatives to reed and sedge were found for thatched roofs, leaving these plants to thrive at the expense of swallowtail caterpillar fare: milk parsley.

084

085

086

087

088

089

090

As a butterfly collector himself, Moses Harris describes in the prelude to *The Aurelian* a method to prevent smaller insects from attacking pinned specimens: "Take the drawer before it be lined with cork and set it some distance from the fire so as to obtain a little warmth, then with a small quan-tity of Unguentum Serulium or ointment of Russet, on a woollen rag, rub it all over, inside and out, pretty strongly…the cork which you propose to line it with, and in covering the inside of your drawer with white paper; on no account make use of paste, as some sorts of those insects which destroy them Flies are very fond of it; but strong Gum Arabick instead."

092

091

Class	INSECTA
Sub-Class	PTERYGOTA
Division	ENDOPTERYGOTA
Order	

093

094

095

096

097

098

099

100

101

102

The butterfly named on this page as *Thersamonolycaena dispar* (*no. 094*) is commonly known as the large copper. This unfortunate sub-species is a lasting monument to the folly of indiscriminate collecting in the absence of ecological knowledge. Once relatively common in the East Anglian Fens of Britain, it was declared extinct ca. 1850. Prior to this, its food plant, the great water dock, became rarer because of continued land drainage. As a result the butterfly also became rarer and therefore, ironically, a much sought-after specimen for collectors, who undoubtedly bear much of the blame for its demise, because the food plant survives to this day. Attempts to reintroduce closely related sub-species from Europe have so far met with failure.

093 *Vanessa io*
094 Large copper (female)—*Thersamonolycaena dispar* ⚘
095 Large copper (male)—*Thersamonolycaena dispar* ⚘
096 Adonis blue—*Polyommatus bellargus* ⚘
097 Small copper—*Lycaena phlaeas* ⚘
098 Scarce copper—*Lycaena virgaurea* ⚘
099 Provence hairstreak—*Tomares ballus* ⚘
100 Painted lady—*Cynthia cardui* ⚘
101 *Polyommatus hiere*
102 Red admiral—*Vanessa atalanta* ⚘

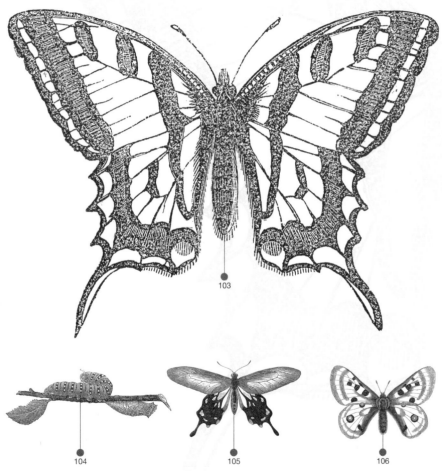

103

104

105

106

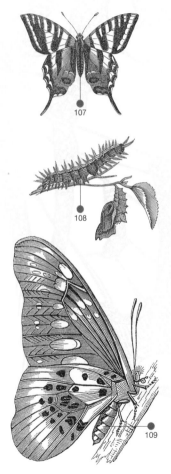

107

108

109

Unlike moths, few butterfly species are serious crop pests. However, the cabbage white butterfly, which originated in Europe, has now been introduced into America and is undoubtedly the most serious butterfly pest on both sides of the Atlantic. Other, more spectacular members of this family (*no. 110*) prefer wild plants and include the yellow brimstone butterfly (*no. 117*)—the species which, because of its color, is often credited with giving butterflies their name. Their close relatives in Europe include the swallowtails and apollos (*no.s 103 and 106*), although they are easily caught by collectors and so among the most endangered.

103 Southern swallowtail—*Papilio alexanor* ✎
104 Common swallowtail caterpillar—*Papilio machaon* ✎
105 *Papilio coon* ✎
106 Apollo butterfly—*Parnassius apollo* ✎
107 American swallowtail—*Papilio ajax* ✎
108 *Argynne paphia*
109 *Satyrus latreille* ✎
110 *Pieris valeria*
111 *Cyllo* sp ✎

110

111

112

113

114

115

116

117

118

Few insects capture our attention as completely as a massive group of migrating butterflies. "Darwin tells us that several times, when the *Beagle* had been some miles...off the shores of Northern Patagonia, the air was filled with insects: that one evening, when the ship was about ten miles from the Bay of San Blas, vast numbers of Butterflies...in flocks of countless myriads extended as far as the eye could range. The seamen cried out 'it was raining butterflies' and such a fact, continues Darwin, was the appearance. Mr Charles Anderson encountered in South-western Africa,...such immense myriads of...butterflies that the sound caused by their wings was such to resemble the 'distant murmuring of waves on a seashore'" (Frank Cowan, 1865).

MOTHS *Lepidoptera*

L ike butterflies, moths have been irresistible to entomologists throughout the ages. Their spectacular markings and colors, coupled with their ease of capture in moth traps, made them favorites among collectors, particularly in the 18th and 19th centuries. Western lepidopterists traveling to the uncharted rain forests of the Old and New Worlds at that time were amazed to find specimens almost as large as their pith helmets. There followed a burgeoning of taxonomic works, and huge collections of moths still exist in museums around the world. The large numbers of pests in this group have attracted the attention of economic entomologists since agriculture began, while silk farming (sericulture) is beaten only by beekeeping as the oldest form of commercial insect rearing.

What is the difference between a butterfly and a moth? This is a frequent, but not easily answered, question because it concerns an artificial division that relates as much to popular folklore as it does to systematic entomology. For example, hawkmoths are more closely related to butterflies than they are to geometrid moths. Another popular fallacy is that moths are night flyers with mainly drab or cryptic coloration, while butterflies are colorful day flyers. However, the beautiful hummingbird hawkmoth flies equally well by day or night; and many of the day-flying butterflies are a drab brown color. All butterflies have club- or knob-shaped antennae; only a few European moths have such structures, and the rest of the world has to rely on the following differences: Butterflies mostly fold their wings together above their bodies when at rest, while moths largely lay them flat, or tentlike over their backs. Many moths have small spines protruding from their hindwings, effectively linking them to the forewings during flight, but butterflies never feature such spines.

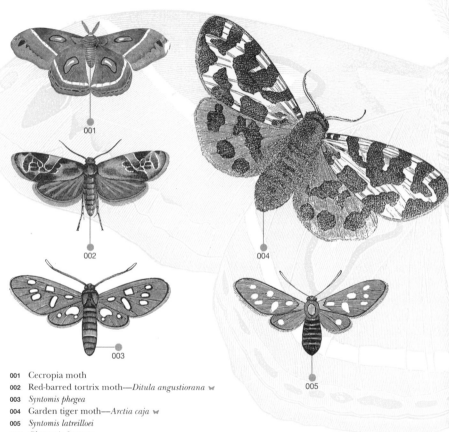

001 Cecropia moth
002 Red-barred tortrix moth—*Ditula angustiorana* ⋙
003 *Syntomis phegea*
004 Garden tiger moth—*Arctia caja* ⋙
005 *Syntomis latreilloei*
006 *Glaucopis formosa*
007 *Glaucopis madagascariensis*
008 *Amphydasis prodromaria*
009 Cinnabar moth—*Callimorpha jacobeae* ⋙

MOTHS *1*

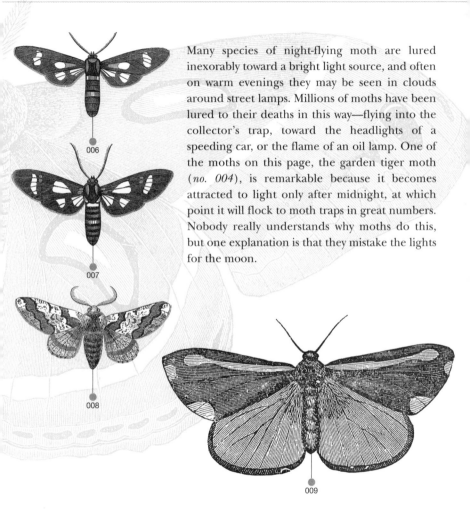

Many species of night-flying moth are lured inexorably toward a bright light source, and often on warm evenings they may be seen in clouds around street lamps. Millions of moths have been lured to their deaths in this way—flying into the collector's trap, toward the headlights of a speeding car, or the flame of an oil lamp. One of the moths on this page, the garden tiger moth (*no. 004*), is remarkable because it becomes attracted to light only after midnight, at which point it will flock to moth traps in great numbers. Nobody really understands why moths do this, but one explanation is that they mistake the lights for the moon.

006

007

008

009

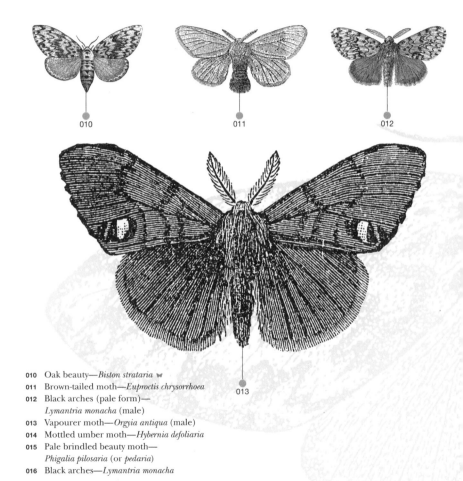

010

011

012

013

010 Oak beauty—*Biston strataria* ⋈
011 Brown-tailed moth—*Euproctis chrysorrhoea*
012 Black arches (pale form)—
      Lymantria monacha (male)
013 Vapourer moth—*Orgyia antiqua* (male)
014 Mottled umber moth—*Hybernia defoliaria*
015 Pale brindled beauty moth—
      Phigalia pilosaria (or *pedaria*)
016 Black arches—*Lymantria monacha*

MOTHS 2

014

015

Processionary moths closely resemble the brown tailed moth (*no. 011*) in that their caterpillars group together when feeding. However, the processionary moth caterpillars march from one food source to another in single file. One fascinating experiment carried out in the United States manipulated the caterpillars so that they formed a closed circle. Amazingly, they followed each other in a circle for several days, stopping only when exhausted. Further experiments revealed that if the circle was deliberately and randomly broken, the caterpillars blindly followed the individual in front. This led to the conclusion that such behavior was instinctive and that no one individual was selected to carry out the role of leader.

016

023

024

When a female moth is ready to lay her eggs, she releases a plume of the chemical pheromone. This is a perfume to which male moths of the same species are specially attuned. Recent studies have found that the large surface area of the male's feathery antennae can detect the female's scent as far as 7 miles/11 km away. Experiments using wind tunnels have also shown that the male can track minute quantities of pheromone and can navigate successfully along the scent plume for many miles by casting from side to side, unerringly finding his mate even when the perfume is being dispersed and eddied by the wind. In just a few species, male moths group together and release pheromones to attract the females.

017 Noctuid moth
018 *Anticlea sinuata*
019 *Cidaria sagittata*
020 *Scotosia certata*
021 *Chesias spartiata*
022 *Tanagra choerophyllata*
023 Winter moth—
    Cheimatobia brumata ✒
024 *Oporabia dilatata*
025 *Melanippe hastata*
026 Beautiful carpet moth—
    Mesoleuca albicillata ✒
027 *Melanippe montana*

025

026

027

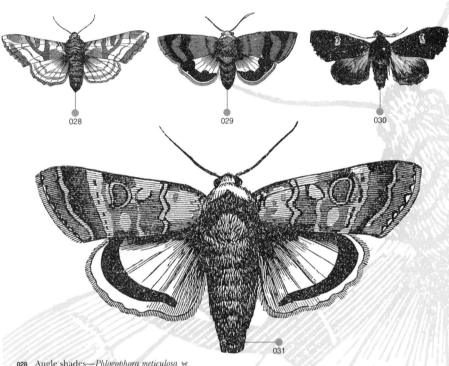

028   Angle shades—*Phlogophora meticulosa* ✎

029   Lesser broad-bordered yellow underwing—
       Noctua ianthina ✎

030   Dot moth—*Melanchra persicariae* ✎

031   Large yellow underwing—*Noctua pronuba* ✎

032   Burnished brass moth—*Diachrisia chrysitis* ✎

033   Herald moth—*Scoliopteryx libatrix* ✎

034   Mother Shipton—*Callistege mi* ✎

035   Red underwing—*Catocala nupia* ✎

036   Clifden nonpareil—*Catocala fraxini* ✎

MOTHS 4

032

033

034

035

Most adult moths feed on nectar from flowers and trees, or by piercing fruit to obtain the juices. Salt is required for reproduction, and many moths have taken to drinking the tears of large mammals such as cattle, often spreading disease as they do so. One species in Asia even uses its proboscis to poke large buffalo in the eye to stimulate tears. The moths on this page feed mainly on nectar or rotting fruits. Bizarrely, one of their cousins from Malaya is the only moth species to have taken the unexpected evolutionary leap of drinking blood. Its proboscis is short, sturdy, and capable of piercing the hides of animals like cattle and antelope.

036

MOTHS 5

The larvae of moths and butterflies (and many other insects that undergo a "complete" metamorphosis) usually eat different food from the adults. Thus evolution occurs simultaneously in two different directions within the life cycle. Most of the moths on this page are from a family in which many of the larvae have elaborate defenses. One of the best-armed caterpillars is that of the puss moth (*no. 044 or 045*), which is able to withdraw its head into its thorax, giving it a grotesque appearance. In addition, its last pair of prolegs have evolved into fleshy "horns," each containing a whiplike filament, which it waves threateningly when disturbed. To complete its armory, it can eject an irritating fluid from a gland in its thorax.

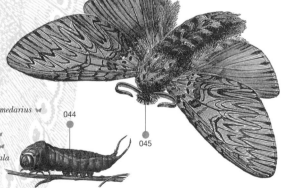

037   Atlas moth
038   Lobster moth—*Stauropus fagi*
039   Iron prominent moth—*Notodonta dromedarius*
040   Buff tip moth—*Phalera bucephala*
041   Sallow kitten moth—*Furcula furcula*
042   Peach blossom moth—*Thyatira batis*
043   Figure-of-8 moth—*Diloba caeruleocephala*
044   Puss moth—*Cerura vinula*
045   Puss moth—*Cerura vinula*

Class	INSECTA
Sub-Class	PTERYGOTA
Division	ENDOPTERYGOTA
Order	

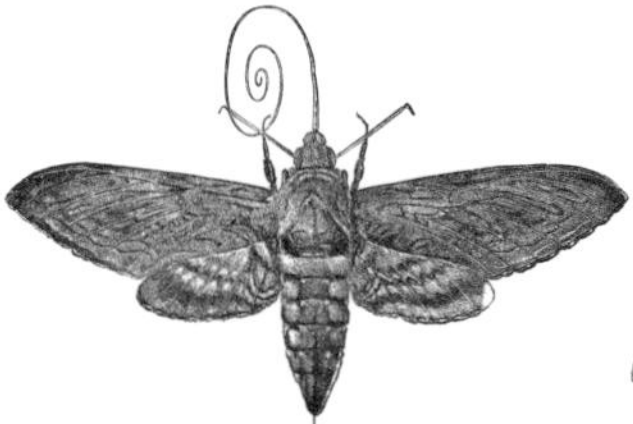

046

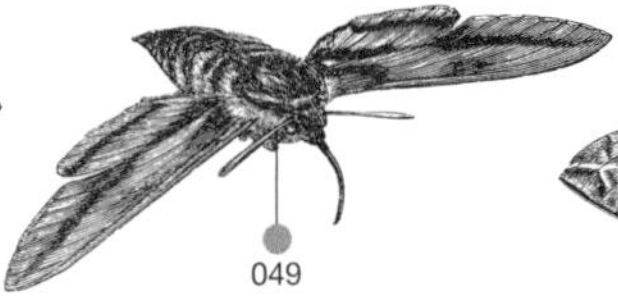

049

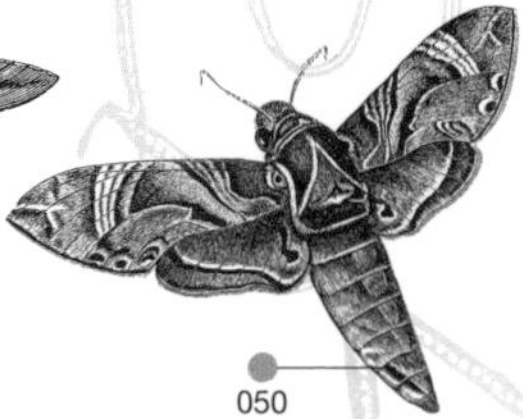

050

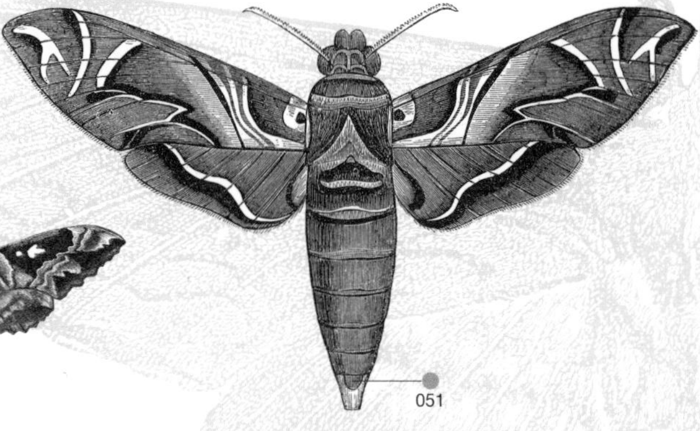

051

047

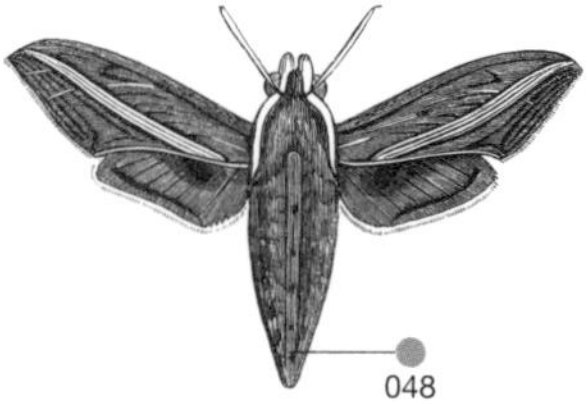

048

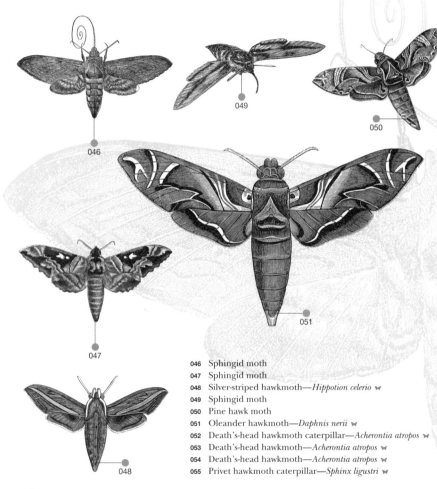

- **046** Sphingid moth
- **047** Sphingid moth
- **048** Silver-striped hawkmoth—*Hippotion celerio*
- **049** Sphingid moth
- **050** Pine hawk moth
- **051** Oleander hawkmoth—*Daphnis nerii*
- **052** Death's-head hawkmoth caterpillar—*Acherontia atropos*
- **053** Death's-head hawkmoth—*Acherontia atropos*
- **054** Death's-head hawkmoth—*Acherontia atropos*
- **055** Privet hawkmoth caterpillar—*Sphinx ligustri*

052

053

"To the superstitious imaginations of the Europeans, the conspicuous markings on the back of the large evening moth, the *Acherontia atropos* [*see no. 053 or 054*], represent the human skull...; hence it is called the death's-head moth, the death's-head phantom...etc. Its cry, which closely resembles the noise caused by the creaking of cork,...certainly more than enough to frighten the ignorant and superstitious, is considered the voice of anguish, the moaning of a child, the signal of grief; and it is regarded 'not as the creation of a benevolent being, but as the device of evil spirits'—spirits, enemies to man, conceived and fabricated in the dark; and the very shining of its eyes is supposed to represent the fiery element whence it is thought to have proceeded" (Frank Cowan, 1865).

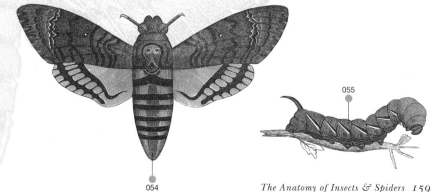

055

054

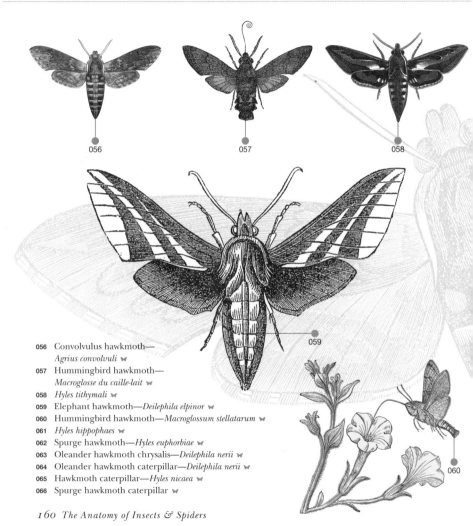

056 Convolvulus hawkmoth—
Agrius convolvuli ✦

057 Hummingbird hawkmoth—
Macroglosse du caille-lait ✦

058 *Hyles tithymali* ✦

059 Elephant hawkmoth—*Deilephila elpinor* ✦

060 Hummingbird hawkmoth—*Macroglossum stellatarum* ✦

061 *Hyles hippophaes* ✦

062 Spurge hawkmoth—*Hyles euphorbiae* ✦

063 Oleander hawkmoth chrysalis—*Deilephila nerii* ✦

064 Oleander hawkmoth caterpillar—*Deilephila nerii* ✦

065 Hawkmoth caterpillar—*Hyles nicaea* ✦

066 Spurge hawkmoth caterpillar ✦

061

062

The beautiful sphinx moths, so called because of the "resemblance between the attitude assumed by the larvae...when disturbed, and that of the Egyptian Sphinx," include among their number the harmless death's-head moth and spectacular hawkmoths. These were favorites among collectors, and the wonder is that so many species survived their onslaught in the late 19th century. Many of these moths remain hovering, as their long tongues probe flowers for nectar. "Most of the 'Hawks' fly by night...but if one will remain perfectly still near a bed of petunias or verbenas, on which fall the rays of a lantern, he will almost certainly have the pleasure of witnessing the wonderful movements, and of hearing the hum of these beautiful insects" (W. Fureaux, 1893).

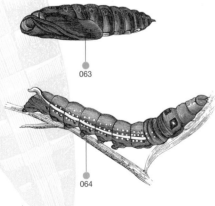

063

065

064

066

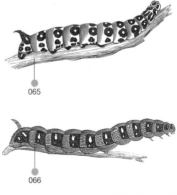

Class INSECTA

Sub-Class PTERYGOTA

Division ENDOPTERYGOTA

Order

067

070

071

072

068

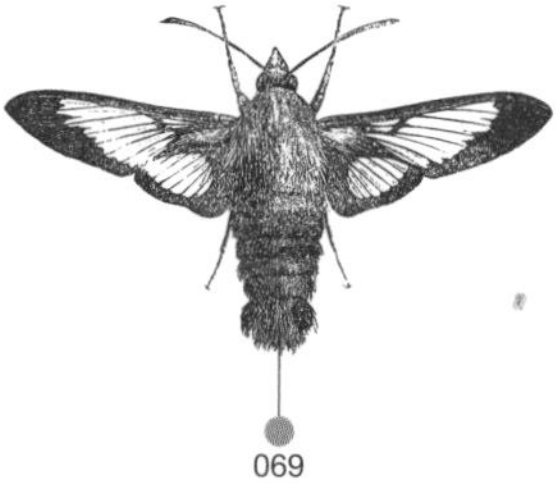

069

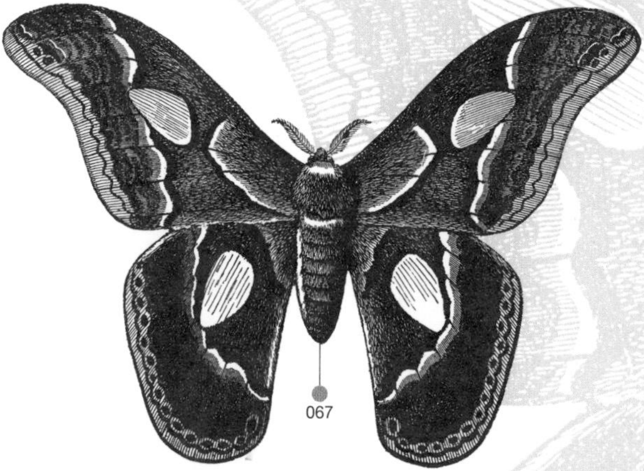

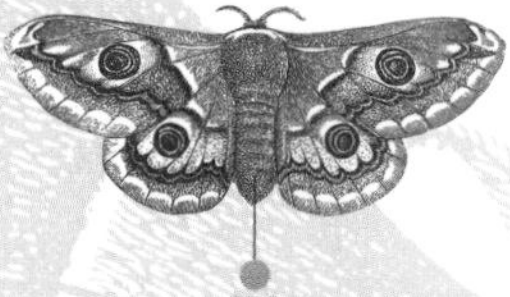

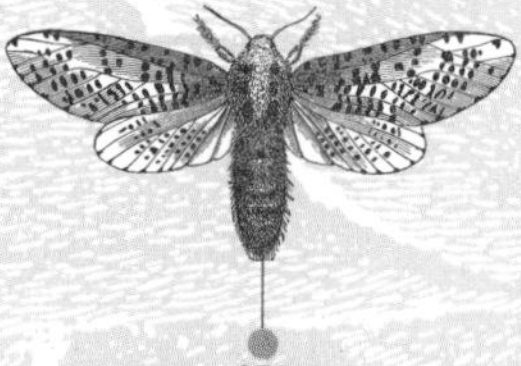

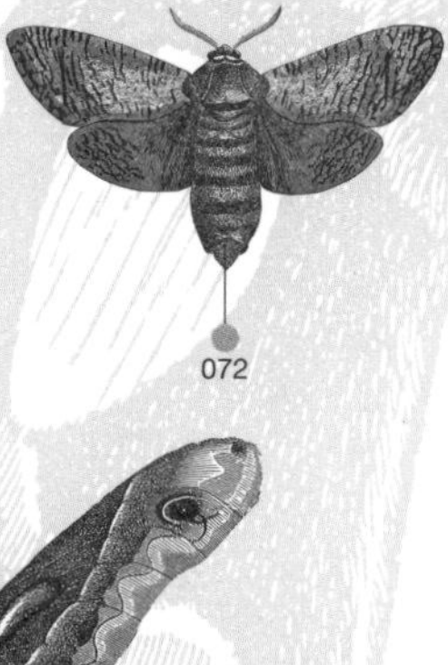

073

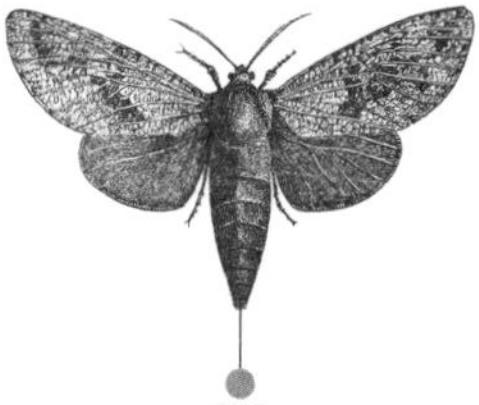

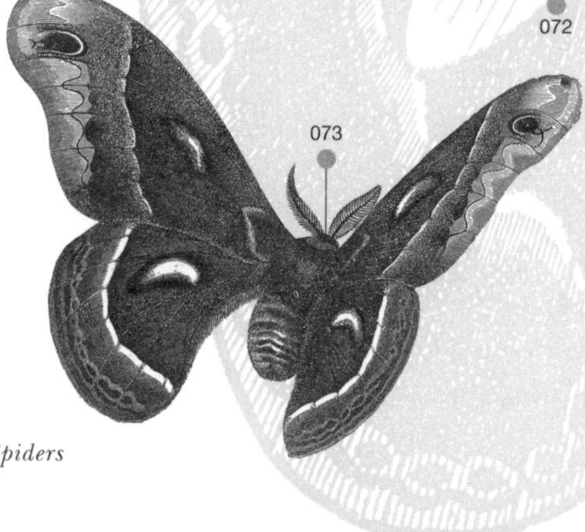

MOTHS 8

The large cecropia moth (*no. 073*) is familiar to many in the United States, but even this is dwarfed by the giant silk moths of India. The cultivated silk moth is a close relative, but, following thousands of years of selective breeding, is no longer found in the wild, even in its native China. Several varieties exist, each capable of producing more than ½ mile /0.8 km of unbroken silk, but none can fly. The value and luxury of silk preoccupied the ancient Greeks and Romans and did much to shape the economic and cultural history of early Europe.

067   Unidentified moth
068   Wood-borer moth
069   Broad-bordered bee hawkmoth—*Hemaris fuciformis*
070   Emperor moth—*Saturnia pavonia*
071   Leopard moth—*Zeuzera pyrina*
072   Goat moth—*Cossus cossus*
073   Cecropia moth
074   Currant clearwing moth larva
075   Currant clearwing moth—*Trochilium tipuliforme*
076   Goat moth—*Cossus cossus*
077   White plume moth—*Pterophorus pentadactyla*
078   Many-plumed moth—*Alucita hexadactyla*

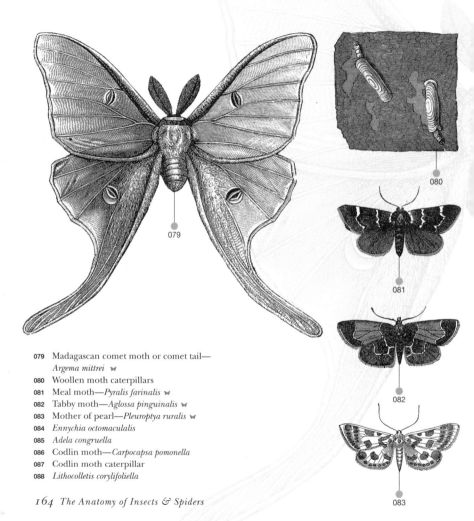

079 Madagascan comet moth or comet tail—
   Argema mittrei 🦋

080 Woollen moth caterpillars

081 Meal moth—*Pyralis farinalis* 🦋

082 Tabby moth—*Aglossa pinguinalis* 🦋

083 Mother of pearl—*Pleuroptya ruralis* 🦋

084 *Ennychia octomaculalis*

085 *Adela congruella*

086 Codlin moth—*Carpocapsa pomonella*

087 Codlin moth caterpillar

088 *Lithocolletis corylifoliella*

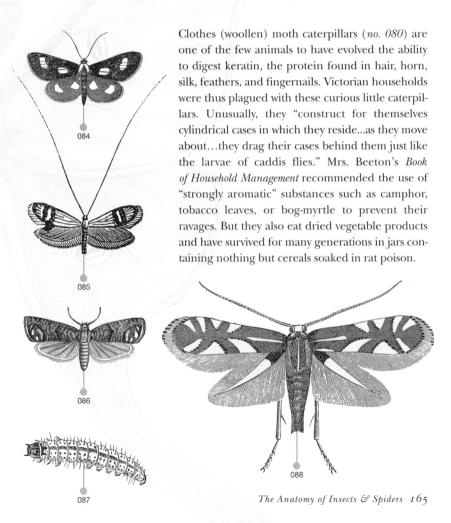

084

085

086

087

088

Clothes (woollen) moth caterpillars (*no. 080*) are one of the few animals to have evolved the ability to digest keratin, the protein found in hair, horn, silk, feathers, and fingernails. Victorian households were thus plagued with these curious little caterpillars. Unusually, they "construct for themselves cylindrical cases in which they reside...as they move about...they drag their cases behind them just like the larvae of caddis flies." Mrs. Beeton's *Book of Household Management* recommended the use of "strongly aromatic" substances such as camphor, tobacco leaves, or bog-myrtle to prevent their ravages. But they also eat dried vegetable products and have survived for many generations in jars containing nothing but cereals soaked in rat poison.

089

090

091

092

093

094

095

MOTHS *10*

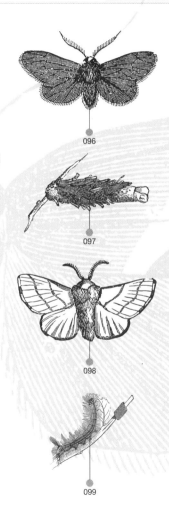

The caterpillars of many moth species are devastating crop pests, variously specializing on roots, seeds, fruits, leaves, flowers, or stems. Some are tiny and tunnel through the leaf or stem tissues, while others are enormous eating machines, devouring large quantities of leaves. Army worms are just such caterpillars, moving en masse and destroying millions of tons of grass in the United States. In Africa a similar species marches through thousands of acres of crop plantations each year, causing fear and famine wherever they occur. In Europe and all other apple-growing regions of the world, moths from the tortrix family render millions of tons of fruits unsaleable by tunneling through the flesh, while their cousins, the pea moths, feed voraciously on young peas.

089 *Epiblema foenella*
090 *Epiblema scutulana*
091 *Xanthosetia zoegana*
092 *Cnephasia octomaculana*
093 *Peronea cristana*
094 *Platepteryx falcula*
095 *Diphthera orion*
096 Bagworm moth—*Psyche opacella*
097 Bagworm moth caterpillar—*Psyche opacella*
098 Lackey moth—*Malacosoma*
099 Lackey moth caterpillar—*Malacosoma*

FLEAS & FLIES
Siphonaptera & Diptera

Confusingly, fleas belong to the sub-class of Pterygota, which means "with wings." But clearly fleas lack wings and their classification is derived from their winged ancestors. Fleas fall into a group called Siphonaptera, after the Greek *siphon* ("tube") and *pteron* ("wing"). The genus name *Pulex* was assigned to fleas by the Romans and is thought to relate to the misconception that fleas originate from dust. In ancient documents, the sweat from slaves' bodies and decomposed matter gave rise to fleas. The distaste with which fleas are regarded is well documented, largely because of their involvement in disease transmission. More agreeable stories, such as the one told by the American entomologist Frank Cowan in *Curious Facts in the History of Insects*, include fleas being sold as pets in Venice.

True or two-winged flies are collectively known as the order Diptera, a vast group of about 90,000 known species. The hindwings are reduced to balancing structures (halteres) and in some species wings are totally absent. Some larvae have economic significance as crop pests, while others attack livestock. The most infamous species are the blood-sucking mosquitoes and tsetse flies, because of the fatalities they can inflict. Yet we must not omit the housefly from concerns about the spread of disease. However, there is some use to be made of in fruit flies. *Drosophila melanogaster*, the vinegar fly, has been the subject of genetic studies for a number of years. It is easy to rear and has a high fecundity and a short life cycle—all factors that make it attractive to work with. The embryologist Thomas Hunt Morgan pioneered genetic studies in inherited traits in the early 20th century, and his work with insects reiterated the work done on plants by Gregor Mendel in the 1860s.

Class	INSECTA		
Sub-Class		PTERYGOTA	
Division			ENDOPTERYGOTA
Order			

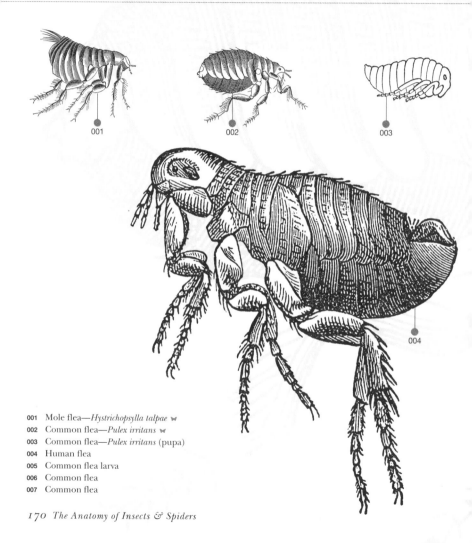

001   Mole flea—*Hystrichopsylla talpae* ⋈

002   Common flea—*Pulex irritans* ⋈

003   Common flea—*Pulex irritans* (pupa)

004   Human flea

005   Common flea larva

006   Common flea

007   Common flea

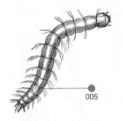

005

006

The ugly Siphonaptera are flattened, which makes them difficult to detach from the hairs and feathers of animals. The irritating common flea has the dangerous ability to transmit disease—murine typhus and bubonic plague being two such flea-carried diseases. The human devastation that fleas caused in China in the 14th century has been well documented, and the bubonic plague (or the Black Death, as it was dubbed in England), which was incorrectly attributed to rats, caused communities from China to Europe to crumble. The peculiarly entertaining flea circuses present a lighter side to these menacing creatures, and touring circuses still exist today.

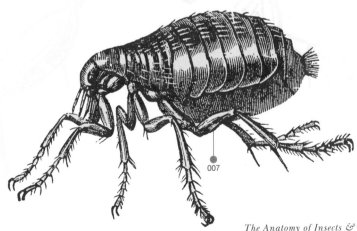

007

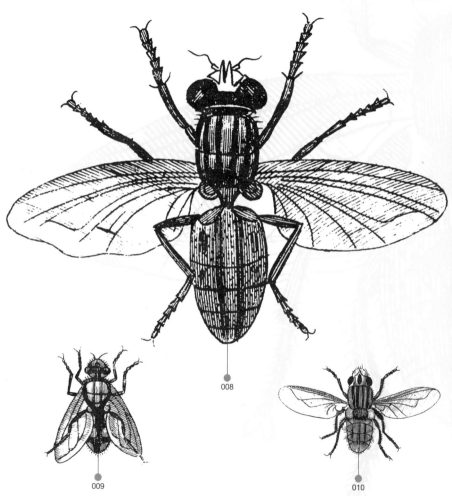

008

009

010

FLIES *1*

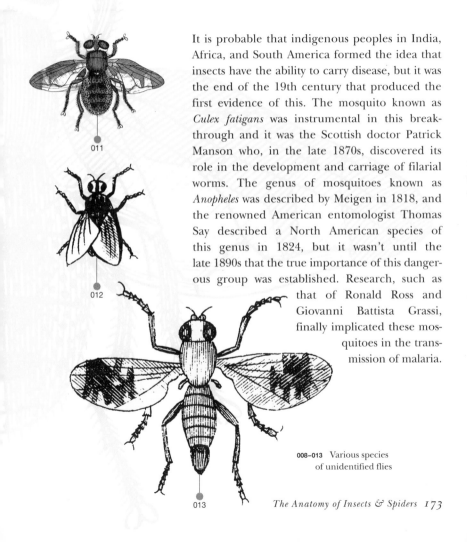

It is probable that indigenous peoples in India, Africa, and South America formed the idea that insects have the ability to carry disease, but it was the end of the 19th century that produced the first evidence of this. The mosquito known as *Culex fatigans* was instrumental in this breakthrough and it was the Scottish doctor Patrick Manson who, in the late 1870s, discovered its role in the development and carriage of filarial worms. The genus of mosquitoes known as *Anopheles* was described by Meigen in 1818, and the renowned American entomologist Thomas Say described a North American species of this genus in 1824, but it wasn't until the late 1890s that the true importance of this dangerous group was established. Research, such as that of Ronald Ross and Giovanni Battista Grassi, finally implicated these mosquitoes in the transmission of malaria.

011

012

008–013 Various species of unidentified flies

013

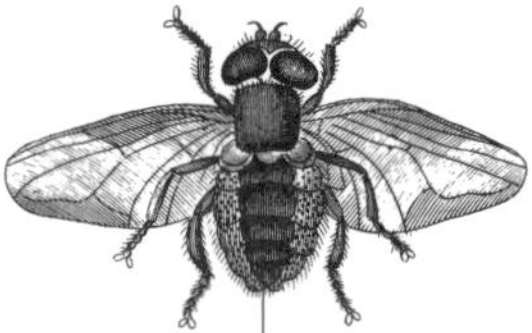

014

015

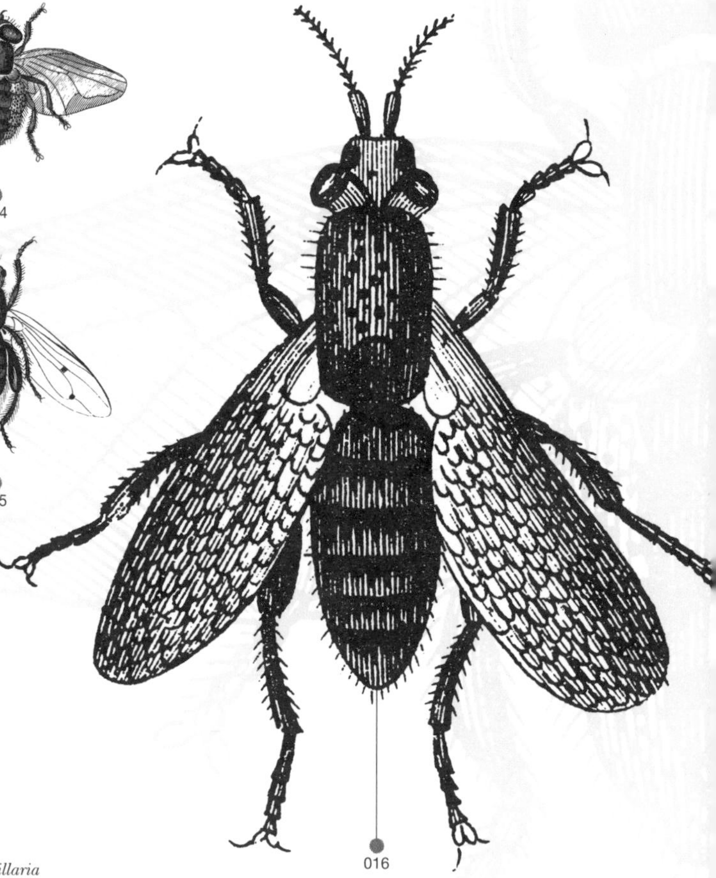

016

014 Unidentified fly
015 Unidentified fly
016 Unidentified fly
017 Unidentified fly
018 Unidentified fly
019 *Ornithomyia fringillaria*
020 Scuttle fly—*Phora abdominalis* ✎

174 The Anatomy of Insects & Spiders

017

018

While medical entomology first implicated insects in the role of biological disease transmission, it also recognized the possibility of mechanical transmission on body parts. Fleas were assumed to be carriers of the bubonic plague; houseflies as carriers of typhoid; and lice as transmitters of typhus fever. Medical entomology was founded and a new and very powerful aspect of this ancient study evolved. The detrimental nature of flies to humans, livestock, and crops means that our emphasis is on the negativity that they bring to our lives. Alternatively, documents show the Native America Indians studied and observed insects as creatures to live beside them. One belief was that flies were thought to teach them to mourn.

019

020

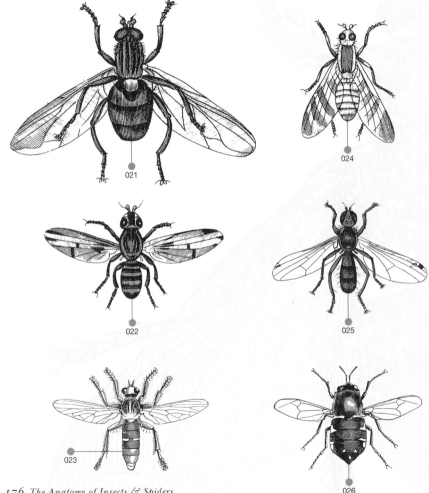

027

028

Minute shore-fly pupae were collected by Native Americans in the Mono Lake region of California for the fat body inside the outer casing. The edible morsels were called *kootsabe* and were eaten raw. History is speckled with accounts of insects being used as a source of food, and this practice remains fully integrated into some cultures today. As well as insects being recognized for their nutritional value to humans, the Chinese transposed this merit to the production of animal feed, and the Cantonese reared fly larvae for medicine, as well as for food.

021 Unidentified fly
022 Unidentified fly
023 Robber fly—*Asilus crabroniformis* ✎
024 Unidentified fly
025 Unidentified fly
026 Soldier fly—*Stratiomys furcata* ✎
027 *Asilus germanicus*
028 Drone fly—*Eristalis tenax* ✎
029 Meat fly

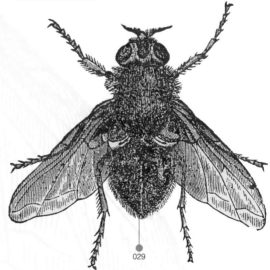

029

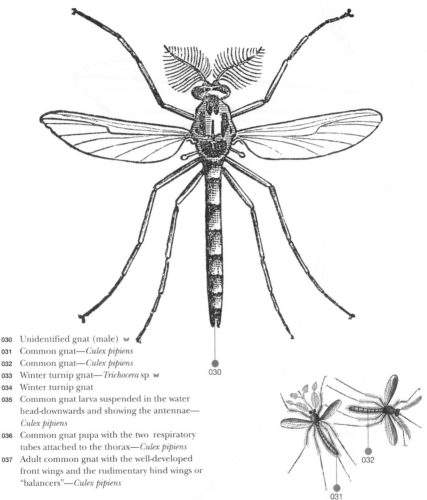

030 Unidentified gnat (male) ✎

031 Common gnat—*Culex pipiens*

032 Common gnat—*Culex pipiens*

033 Winter turnip gnat—*Trichocera* sp ✎

034 Winter turnip gnat

035 Common gnat larva suspended in the water head-downwards and showing the antennae—*Culex pipiens*

036 Common gnat pupa with the two respiratory tubes attached to the thorax—*Culex pipiens*

037 Adult common gnat with the well-developed front wings and the rudimentary hind wings or "balancers"—*Culex pipiens*

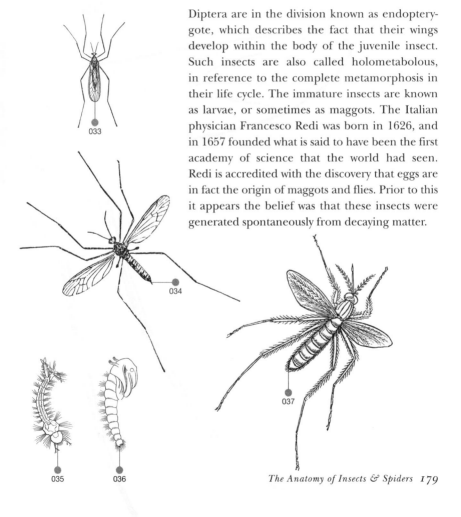

Diptera are in the division known as endopterygote, which describes the fact that their wings develop within the body of the juvenile insect. Such insects are also called holometabolous, in reference to the complete metamorphosis in their life cycle. The immature insects are known as larvae, or sometimes as maggots. The Italian physician Francesco Redi was born in 1626, and in 1657 founded what is said to have been the first academy of science that the world had seen. Redi is accredited with the discovery that eggs are in fact the origin of maggots and flies. Prior to this it appears the belief was that these insects were generated spontaneously from decaying matter.

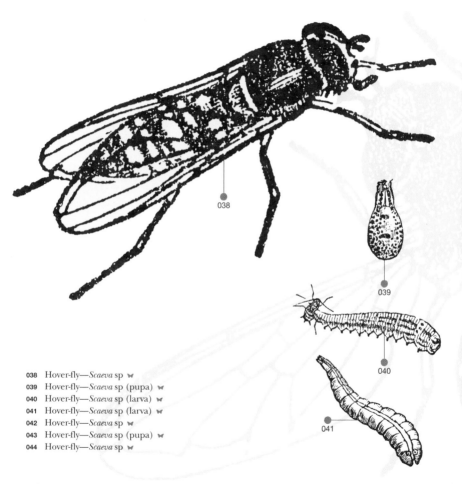

038 Hover-fly—*Scaeva* sp ￦

039 Hover-fly—*Scaeva* sp (pupa) ￦

040 Hover-fly—*Scaeva* sp (larva) ￦

041 Hover-fly—*Scaeva* sp (larva) ￦

042 Hover-fly—*Scaeva* sp ￦

043 Hover-fly—*Scaeva* sp (pupa) ￦

044 Hover-fly—*Scaeva* sp ￦

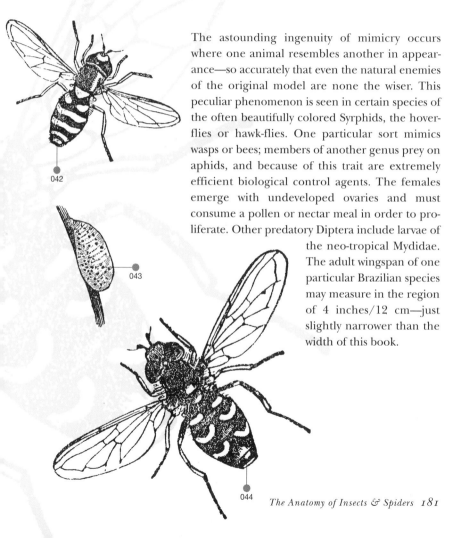

042

043

The astounding ingenuity of mimicry occurs where one animal resembles another in appearance—so accurately that even the natural enemies of the original model are none the wiser. This peculiar phenomenon is seen in certain species of the often beautifully colored Syrphids, the hover-flies or hawk-flies. One particular sort mimics wasps or bees; members of another genus prey on aphids, and because of this trait are extremely efficient biological control agents. The females emerge with undeveloped ovaries and must consume a pollen or nectar meal in order to proliferate. Other predatory Diptera include larvae of the neo-tropical Mydidae. The adult wingspan of one particular Brazilian species may measure in the region of 4 inches/12 cm—just slightly narrower than the width of this book.

044

Class INSECTA

Sub-Class PTERYGOTA

Division ENDOPTERYGOTA

Order

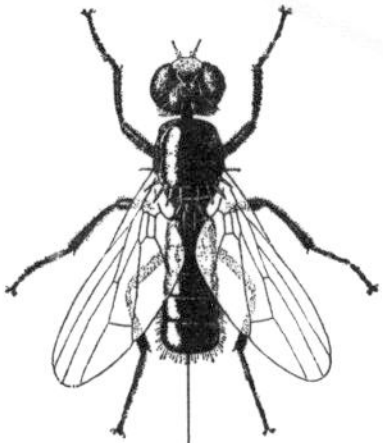

045

046

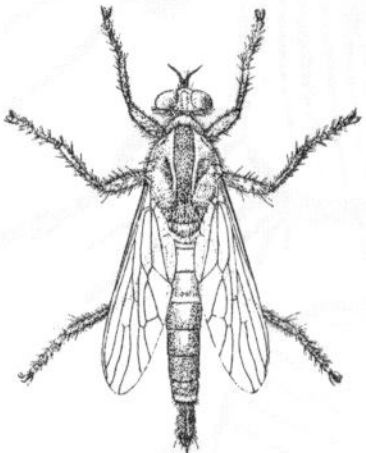

047

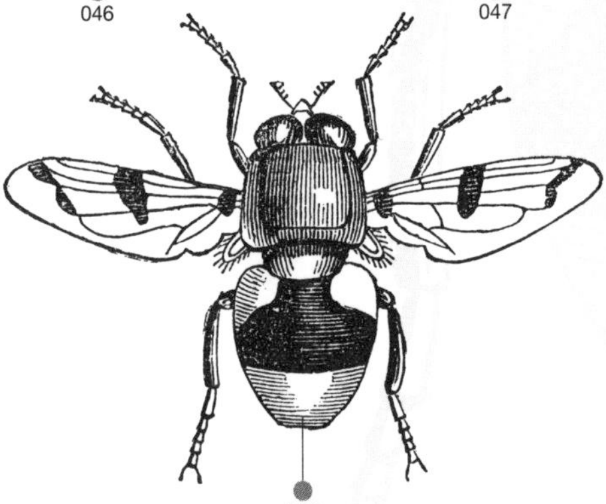

048

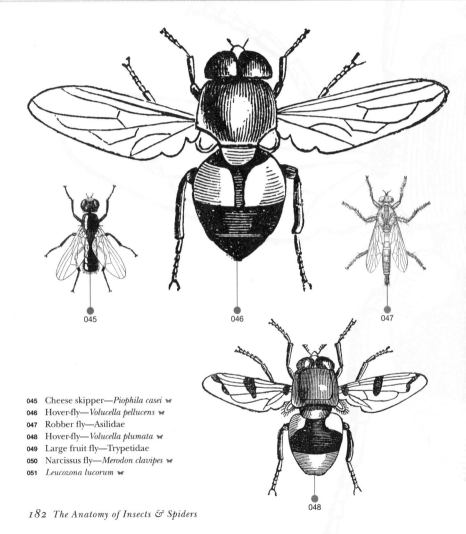

045 Cheese skipper—*Piophila casei*
046 Hover-fly—*Volucella pellucens*
047 Robber fly—Asilidae
048 Hover-fly—*Volucella plumata*
049 Large fruit fly—Trypetidae
050 Narcissus fly—*Merodon clavipes*
051 *Leucozona lucorum*

"A fly, Sir, may sting a stately horse and make him wince; but one is but an insect, and the other is a horse still."
SAMUEL JOHNSON

Among the fascinating anecdotes surrounding the often unpopular flies are those that are straight-forward accounts of their force in numbers. And a most charming story is retold by the Roman writer Pliny: The curious ingredient of fly heads, on occasion mixed with paper ashes, nuts, breast milk, or colewort, was reported to be efficacious in the cure of baldness. Other anecdotes tell of immortal flies in Scotland, brass flies in Naples, and of ancient Greek frivolities incorporating flies—all presenting a lighter facet to the group we call the Diptera. Venerable tricks to combat marauding flies include daubs of lion fat, wolf tails entombed in the afflicted building, or—less dangerously accomplished—bait traps of herbs, suspended like conventional flypaper.

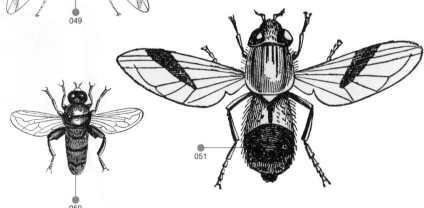

049

050

051

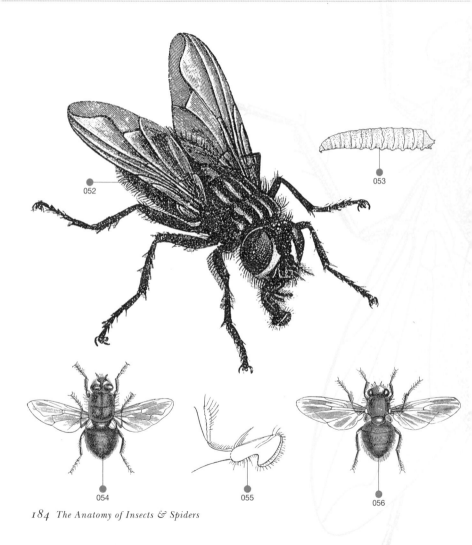

057

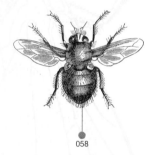

058

The widely disliked housefly (*no. 052*) has a world-wide distribution. It is connected with uncleanliness because of its common association with ordure and its equally common attraction to our food. Germ transmission is also caused through its frequent defecation. The medley of diseases that may be spread by the housefly includes typhoid, diarrhea, anthrax, amebic and bacillary dysentery, poliomyelitis, tuberculosis, and cholera. Indeed, Frank Cowan documents in his 1865 book an assumption that flies died of the last disease. Flies also act as intermediate hosts of roundworms and can carry tapeworms. Disease was said to be imminent by the ancient Egyptians, Persians, and Indians if flies flew into their dreams.

052 Common housefly—*Musca domestica* 🦟
053 housefly (larva)—*Musca* sp
054 *Musca chloris* (male)
055 *Musca chloris* (tongue)
056 *Musca* or *Anthomyia lardaria* (female)
057 *Tachina fera*
058 *Tachina grossa*
059 Common housefly—*Musca domestica* 🦟

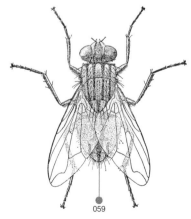

059

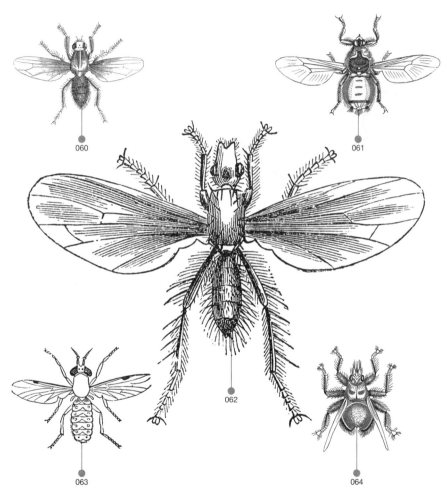

060

061

062

063

064

Botfly is the common name for members of the Gasterophilidae family, although this term has crept into other families. Larvae of these flies are internal parasites of equines—the body of the host being the favored environment for laying eggs in. The larvae migrate to the horse's mouth and, as shown here (*no. 066*), eventually cling to the stomach lining utilizing their mouth hooks. Bright-yellow dung fly males cause a striking contrast to the drab dung that they frequent, while the robust horseflies can be the bane of humans and may administer a nasty bite. Members of this family are often called "stouts." As commonly occurs, it is the ferocious female who is the blood imbiber, rather than the nectar-feeding male.

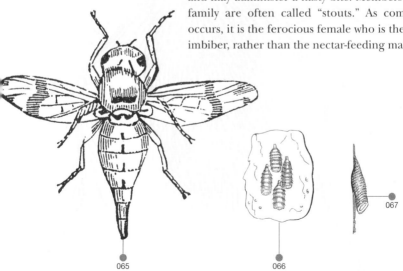

065

066

067

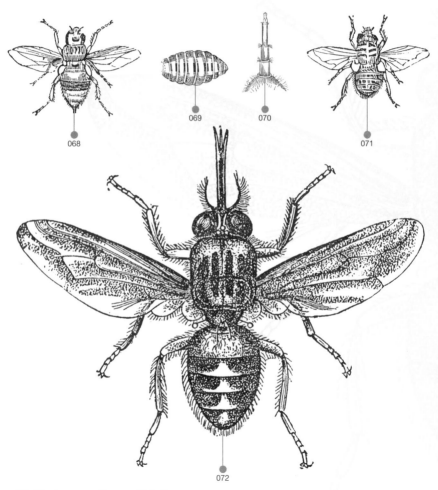

068

069

070

071

072

068 Cattle botfly—*Oestrus bovis*
069 Cattle botfly (larva)—*Oestrus bovis*
070 Cattle botfly (ovipositor, with tubes
     like a telescope—*Oestrus bovis*
071 Sheep botfly—*Oestrus bovis*
072 Tsetse fly—*Glossina morsitans morsitans*
073 Sheep ked—*Melophagus ovinus*
074 *Nycteribia latreille*

The fearsome botflies are the bane of cattle and sheep (*no.s 068–071*). The sheep bot is also commonly known as the sheep nostrilfly, given that this area is the site of its egg lay (oviposition). A misguided ancient use of the burrowing juvenile of this fly was as a remedy for epilepsy. *Myia* is Greek for "fly," and derived from this is the term myiasis, whereby fly larvae are able to develop on living flesh. And the warble flies were historically known as wormuls, wormals, warbles, and bots.

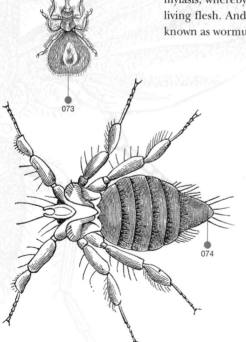

073

074

"*For the Indians, Persians and Egyptians do teach, that if Flies appear to us in our sleep, it doth signifie an herauld at arms, or an approaching disease.*"
FRANK COWAN

Class	INSECTA		
Sub-Class		PTERYGOTA	
Division			ENDOPTERYGOTA
Order			

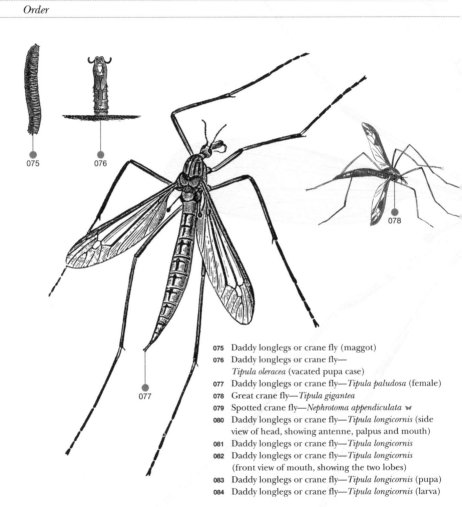

075   Daddy longlegs or crane fly (maggot)
076   Daddy longlegs or crane fly—
      Tipula oleracea (vacated pupa case)
077   Daddy longlegs or crane fly—*Tipula paludosa* (female)
078   Great crane fly—*Tipula gigantea*
079   Spotted crane fly—*Nephrotoma appendiculata* ⚭
080   Daddy longlegs or crane fly—*Tipula longicornis* (side
      view of head, showing antenne, palpus and mouth)
081   Daddy longlegs or crane fly—*Tipula longicornis*
082   Daddy longlegs or crane fly—*Tipula longicornis*
      (front view of mouth, showing the two lobes)
083   Daddy longlegs or crane fly—*Tipula longicornis* (pupa)
084   Daddy longlegs or crane fly—*Tipula longicornis* (larva)

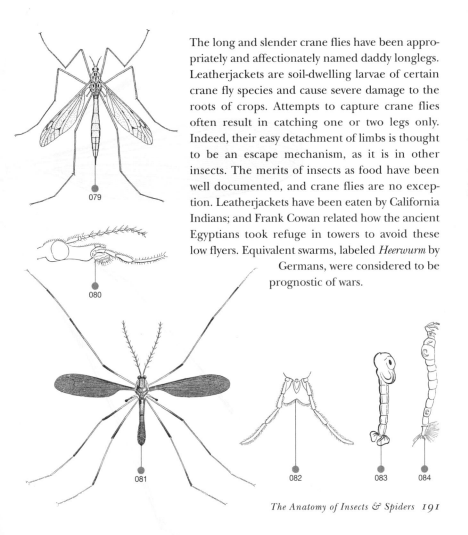

The long and slender crane flies have been appropriately and affectionately named daddy longlegs. Leatherjackets are soil-dwelling larvae of certain crane fly species and cause severe damage to the roots of crops. Attempts to capture crane flies often result in catching one or two legs only. Indeed, their easy detachment of limbs is thought to be an escape mechanism, as it is in other insects. The merits of insects as food have been well documented, and crane flies are no exception. Leatherjackets have been eaten by California Indians; and Frank Cowan related how the ancient Egyptians took refuge in towers to avoid these low flyers. Equivalent swarms, labeled *Heerwurm* by Germans, were considered to be prognostic of wars.

079

080

081

082

083

084

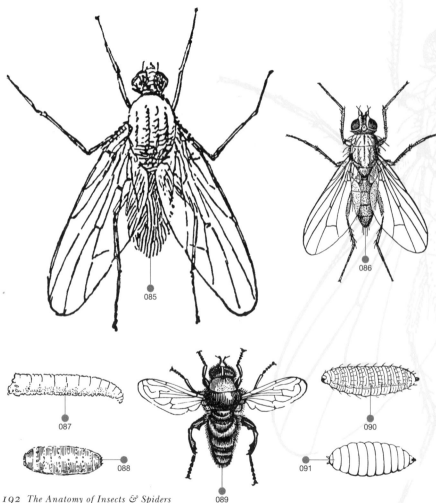

092

Insect pests in the time of the ancient Greeks and as far back as the Pharaohs are, biblically speaking, acknowledged. However, insect damage (at least to crops) warranted little concern (or rather too little to document) in Europe until the early 18th century, and even then the focus was largely on forest pests. The leading naturalist Carolus Linnaeus published material on economic entomology (pest control) in the early part of the 18th century. His book advised farmers on the handling of those insects that threatened their livelihoods. A number of similar publications followed, including Vincent Köhler's volume, *Insects Injurious to Gardeners, Foresters and Farmers,* in the 1830s.

085 Beet fly—*Anthomyia betae*
086 Onion fly—
      Anthomyia ceparum (female)
087 Beet fly—*Anthomyia betae* (larva)
088 Beet fly—*Anthomyia betae* (chrysalis)
089 Narcissus fly—*Merodon equestris*
090 Narcissus fly—
      Merodon equestris (grub)
091 Narcissus fly—
      Merodon equestris (pupa)
092 Carrot fly (larvae appearing from galleries excavated in the carrot—
      Psila rosae
093 Carrot fly—*Psila rosae* (larva)
094 Carrot fly—*Psila rosae* (pupa)
095 Carrot fly—*Psila rosae*

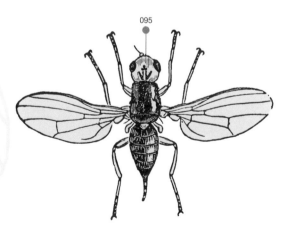

095

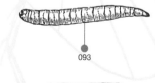

093

094

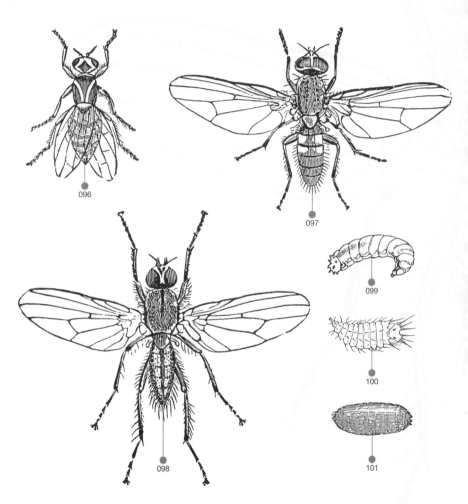

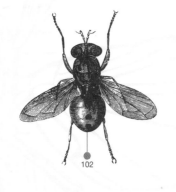

102

Simple common names are assigned to frequent insect pests of agricultural crops on this page. The narcissus fly (*no. 102*) is also known as the bulb fly, both names pertaining to the fact that the larvae feed on plant bulbs. Interestingly, the adult is a mimic of bumblebees—a phenomenon that we have touched on previously and will consider again. It is said that the name of the wheat-stem fly, *Chlorops pumilionis* (*no. 096*), indicates a dwarf, referring to an attacked plant seldom attaining its true height. The wheat-bulb fly was the subject of the first seed treatments to combat pests in the 1950s. As with many pests, the cabbage-root fly has a worldwide distribution intimately connected to its host plants, the brassicas.

096 Wheat-stem fly—*Chlorops pumilionis*
097 Potato fly—*Anthomyia tuberosa* ⋈
098 Radish fly or turnip-root fly—
    Anthomyia radicum ⋈
099 Cabbage fly (larva)—*Anthomyia brassicae*
100 Potato fly—*Anthomyia tuberosa* (larva)
101 Cabbage fly (pupa)—*Anthomyia brassicae*
102 Narcissus fly—*Merodon equestris*
103 Onion fly—*Anthomyia ceparum*

103

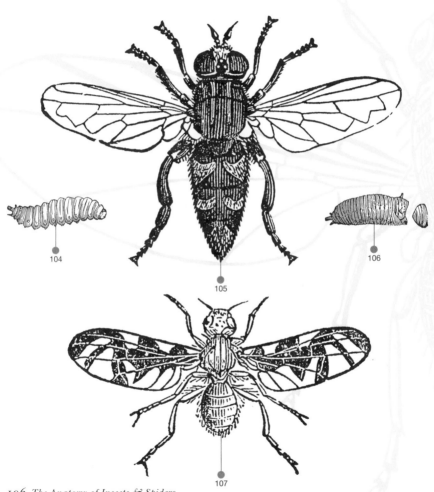

104

105

106

107

To name the Hessian fly is to nominate but one injurious pest whose exploits have regularly littered history (numerous destructive crop pests are shown on this page). Thomas Say documented the Hessian fly's activities as far back as the early part of the 19th century. In fact, 19th-century literature is inundated with papers pertaining to this fly, this no doubt being (in part at least) a reflection of its increasing pest status. It is known as *Mayetiola destructor* and is a member of the Cecidomyiidae family, commonly called the gall midges as a group. However, the Hessian fly is not a gall inducer, but it is a pest in Europe and North America, causing considerable damage to wheat.

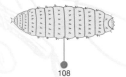

108

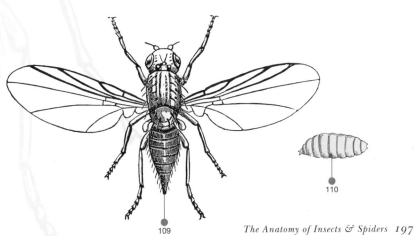

109

110

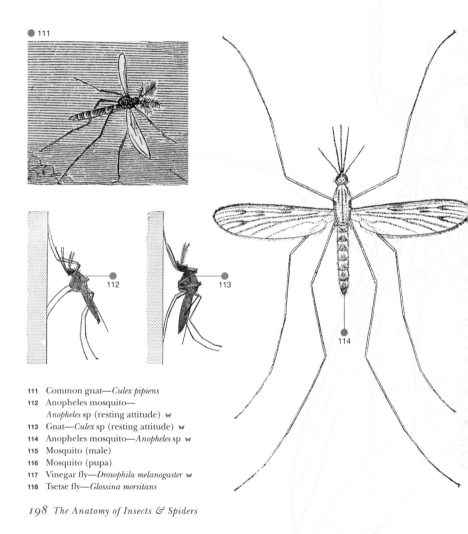

111 Common gnat—*Culex pipiens*
112 Anopheles mosquito—
Anopheles sp (resting attitude) ✎
113 Gnat—*Culex* sp (resting attitude) ✎
114 Anopheles mosquito—*Anopheles* sp ✎
115 Mosquito (male)
116 Mosquito (pupa)
117 Vinegar fly—*Drosophila melanogaster* ✎
118 Tsetse fly—*Glossina morsitans*

Blood-feeding tsetse flies (*Glossina, no. 118*) are medically significant in Africa due to their ability to transmit sleeping sickness to humans and the parasitic disease nagana to cattle. As a consequence of their role as disease carriers, tsetse flies have attracted much attention. It is possible that sleeping sickness first struck in the 14th century, but it was not until the early 20th century that the causal parasites, *Trypanosoma*, were identified and tsetse flies were implicated as the disease vectors.

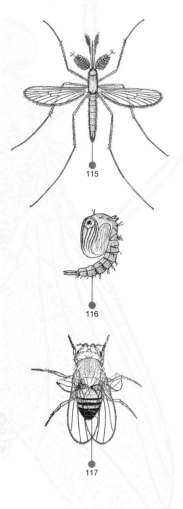

115

116

117

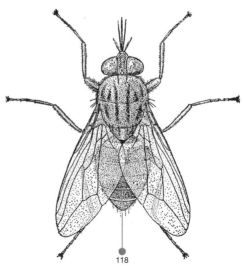

118

GRASSHOPPERS & CRICKETS *Orthoptera*

The Orthoptera are named from the Greek *orthos* (meaning "straight" or "rigid") and *pteron* (meaning "wing"). This refers to the long, almost parallel front wings that most of the adult insects in this order possess. Just a few species of Orthoptera remain wingless throughout their lives.

Orthoptera are familiar insects, including among their number grasshoppers, locusts, ground hoppers, crickets, bush crickets, or katydids, mole crickets, and camel crickets. They have often been involved in events of catastrophic proportions and thus, more than any other group of insects, have attracted the attention of economic entomologists (pest controllers) down the ages. Plagues of locusts have regularly swept across the continents of the world, destroying crops on such a scale that severe famines have often resulted.

From China and Japan to India, the Middle East, Africa, Europe, and the Americas, the great civilizations have battled with this threat, and tried numerous methods for precaution.

Some of the earliest references to locusts appear in Hebrew records dating from the 9th century BC, and the Hebrews, Syrians, Chinese, and ancient Greeks had laws that variously demanded the extermination of eggs, young, and adults. In the Talmud, Jews were even given advice on how to catch them on the Sabbath without breaking the law. Ancient Chinese entomologists referred to the Asian locust as *Fu Chung* and kept amazing records of their plagues from 707 BC right down to the present day. These records have been of immense value, allowing 17th-century Chinese entomologists to link them to the flood or drought cycle of certain river deltas, thereby giving them the ability to accurately predict swarms and to suggest public engineering works on the rivers, which might prevent such plagues occurring.

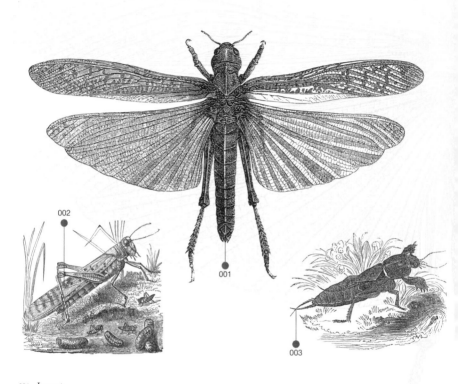

001

002

003

004

001 Locust
002 Migratory locust—*Locusta migratoria* ✎
003 Mole cricket
004 Grasshopper (adult)
005 Unidentified
006 Unidentified
007 Grasshopper
008 Grasshopper

GRASSHOPPERS
ORTHOPTERA **& CRICKETS** *1*

The ancestors of our present-day grasshoppers, locusts, and crickets are believed to have arisen in the great carboniferous forests of more than 300 million years ago. Today there are more than 20,000 species distributed on all the continents of the earth except Antarctica. The smallest found so far is a tiny species from South Africa, no more than ⅟₂₅ inch/1–2 mm in length. The largest individuals are from a species that is toasted and eaten as a delicacy by the Yukpa-Yuko natives of Venezuela and Colombia (known to them as the *sakáramo*). These insects would have problems finding enough room to stand on the palm of your hand!

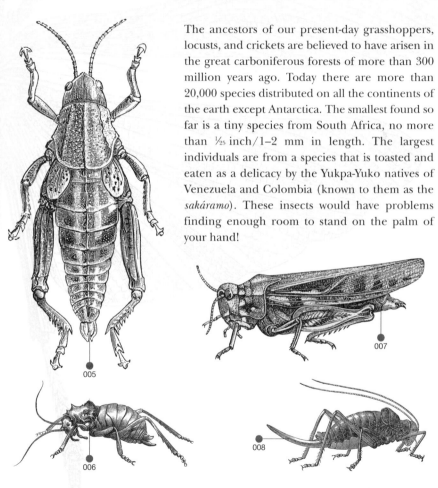

005

007

006

008

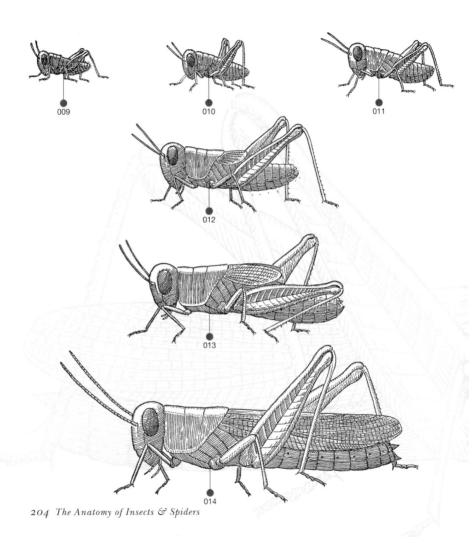

009

010

011

012

013

014

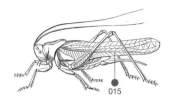

015

009 Grasshopper (1st instar) ✎
010 Grasshopper (2nd instar) ✎
011 Grasshopper (3rd instar) ✎
012 Grasshopper (4th instar) ✎
013 Grasshopper (5th instar) ✎
014 Grasshopper (adult) ✎
015 Grasshopper (adult or imago)
016 Common mole cricket—
     Gryllotalpa vulgaris (1st instar) ✎
017 Common mole cricket eggs—
     Gryllotalpa vulgaris ✎
018 Common mole cricket—
     Gryllotalpa vulgaris (2nd instar) ✎
019 Common mole cricket—
     Gryllotalpa vulgaris (adult)

The Orthoptera and the remaining insects in this book belong to a division of insects known as Exopterygota, in which the wings develop externally and there is no change in shape during the life cycle. Instead, the young insects usually look like smaller versions of the adults and are called nymphs. There may be as many as 15 immature stages in some cricket species. Functional wings are present only in the adult insects; entomologists have somewhat misleadingly called this "incomplete metamorphosis," and it fooled the ancient Egyptians into thinking that immature insects were actually different species. Chinese and Japanese entomologists rarely made such mistakes, and their knowledge of insect metamorphosis was well documented by the 11th century. In the West it was the 17th century before the Italian physician and poet Francesco Redi published the first accurate accounts of insect metamorphosis.

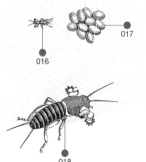

016

017

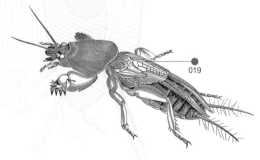

018

019

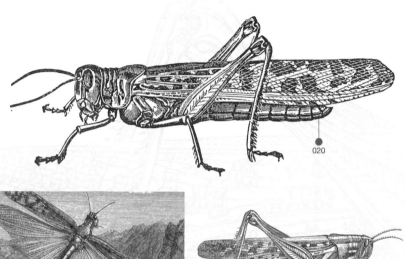

020

021

022

020 Locust
021 Locust
022 Locusts and grasshoppers
023 Locust
024 Grasshopper nymph ⚭
025 Migratory locust—*Locusta migratoria* ⚭

GRASSHOPPERS
<small>ORTHOPTERA</small> & CRICKETS 3

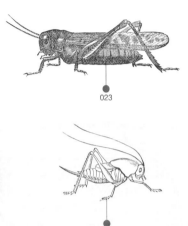

023

024

On locusts: "A day of darkness and of gloomi-ness...as the morning spread upon the mountains....A fire devoureth before them; and behind them a flame burneth: the land is as the garden of Eden before them and behind them a desolate wilderness; yea, and nothing shall escape them... Like the noise of chariots on the tops of mountains shall they leap, like the noise of flame that devoureth the stubble, as a strong people set in battle array....They shall run like mighty men; they shall climb the wall like men of war; and they shall march every one on his ways and they shall not break their ranks....The earth shall quake before them; the heaven shall tremble: the sun and the moon shall be dark, and the stars withdraw their shining"(Joel 2: 2–10).

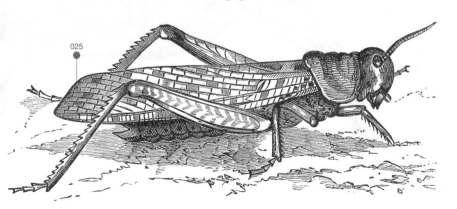

025

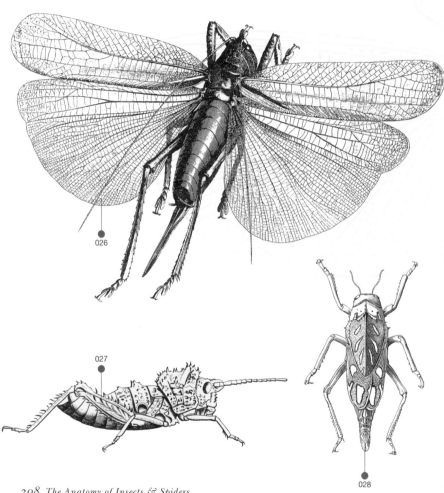

026

027

028

029

Since they first took to the plough 7–12,000 years ago, the agricultural civilizations of the Old and the New Worlds have lived in dread of locust plagues, often seeing them as divine retribution for sins against their god or gods. In Europe in the Middle Ages several lawsuits were brought against them, and there are numerous records of priests excommunicating them. However, to the hunter-gathering peoples of the world, grasshopper and locust swarms are manna from heaven, providing a welcome bounty of protein-rich food. Today entomologists are still more preoccupied with their control than with their esthetic beauty. Locusts have the curious ability to switch between a solitary existence (when population levels are low) and a swarming, social type (when they are more numerous). This change in behavior may be accompanied by changes in color or body shape.

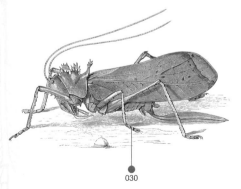

030

026   Cricket
027   *Xiphocerca asina*
028   *Pneumora scutellaris* (female)
029   *Schizodactylus monstrosus* (male)
030   *Eumegalodon blanchardi* (female)

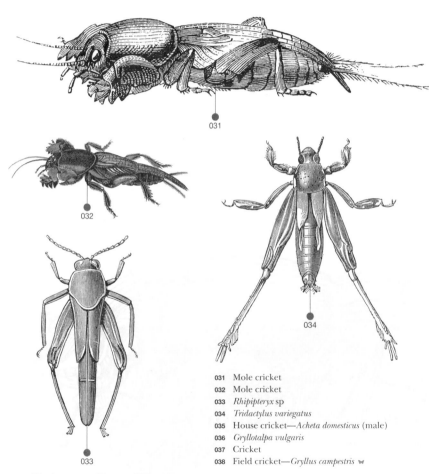

031 Mole cricket
032 Mole cricket
033 *Rhipipteryx* sp
034 *Tridactylus variegatus*
035 House cricket—*Acheta domesticus* (male)
036 *Gryllotalpa vulgaris*
037 Cricket
038 Field cricket—*Gryllus campestris* ⬥

GRASSHOPPERS & CRICKETS 5

ORTHOPTERA

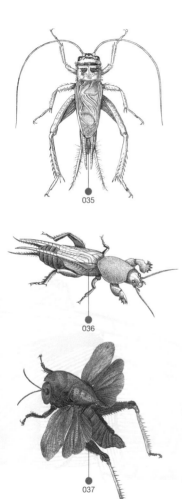

035

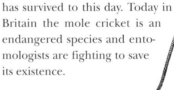

036

The Chinese people have had a special affinity with the cricket for more than 2,000 years. During that time there have been three distinct eras of Cricket Culture. In the Chun Qui period (770–476 BC) they were kept as good-luck charms and symbols of "auspicious virtue." During the Tang dynasty (AD 618–907) and up to the early years of the Song dynasty, male crickets were kept for their beautiful songs, made by rasping their wings together; female crickets "hear" the males using a special organ on their front legs. And during the Song dynasty (AD 960–1279) cricket fighting became a popular national sport that has survived to this day. Today in Britain the mole cricket is an endangered species and entomologists are fighting to save its existence.

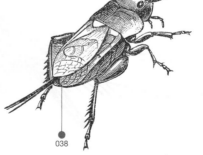

037

038

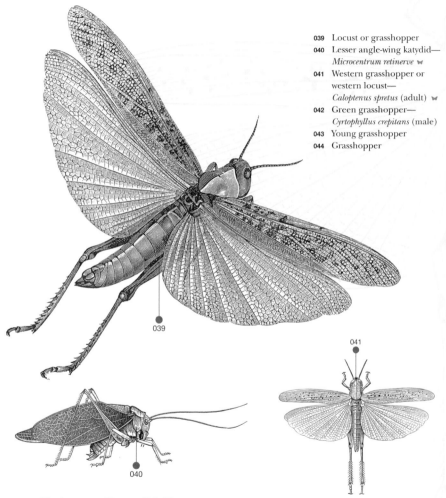

039 Locust or grasshopper
040 Lesser angle-wing katydid—
     Microcentrum retinerve ⚭
041 Western grasshopper or
     western locust—
     Caloptenus spretus (adult) ⚭
042 Green grasshopper—
     Cyrtophyllus crepitans (male)
043 Young grasshopper
044 Grasshopper

GRASSHOPPERS & CRICKETS 6

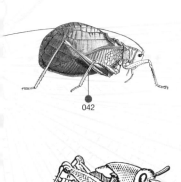

042

043

Grasshoppers differ from most crickets and bush crickets by having much shorter antennae; by "singing" by means of rasping their legs on their wings; and by having their "ears" on the abdomen, instead of the front legs. In crickets, "singing" is the domain of the males, but female grasshoppers may also sing, albeit more softly than the males. Although this group includes locusts and some other agricultural pests, many species are harmless, and the Talmud tells us that they were favorite pets of Jewish children in biblical times. Some species were kosher, leading to an interest in identification among the early Jewish people, who accurately predicted the total number of grasshopper species found in the New World 1,500 years before the first Western entomologists.

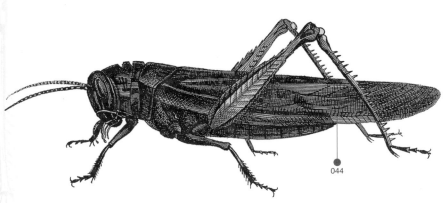

044

STICK INSECTS, COCKROACHES, & PRAYING MANTISES *Phasmida & Dictyoptera*

Early entomologists understandably included stick insects, cockroaches, and praying mantises in the same group as grasshoppers and crickets. Modern entomologists have, however, separated them into two further orders. Stick insects are sufficiently different to warrant their own group, called Phasmids. There are more than 2,000 species of them worldwide, mostly in the tropics. Much beloved as pets, stick insects are generally wingless, although a few, known as "walking leaves," have beautiful netted wings developed into leaflike patterns, which imitate the leaves upon which they live.

Cockroaches and mantises now comprise the order of Dictyoptera, arising from two Greek words meaning "netlike" and "wings." This describes the beautiful netlike wings both groups of insects possess. The first cockroach fossils date from the great forests of the Upper Carboniferous period about 300 million years ago and show that these now-extinct species possessed even more spectacular wings than those of today. There are 3,500 mainly tropical species, most of which do not harbor disease and thrive only in clean, virgin forest. The name *mantis* comes from ancient Greek and means "diviner," but was also used in one of the Idylls of Theocritus (ca. 250 BC) to describe a slim young woman with slender, elongated arms. This describes the shape of many mantis species, but the reference to the diviner may have its origins in the fact that Arab and Turkish cultures believed that these insects constantly prayed with their faces turned toward Mecca. Like grasshoppers and crickets, cockroaches, mantises, and stick insects undergo an incomplete metamorphosis, with the young nymphs following a similar lifestyle to their parents.

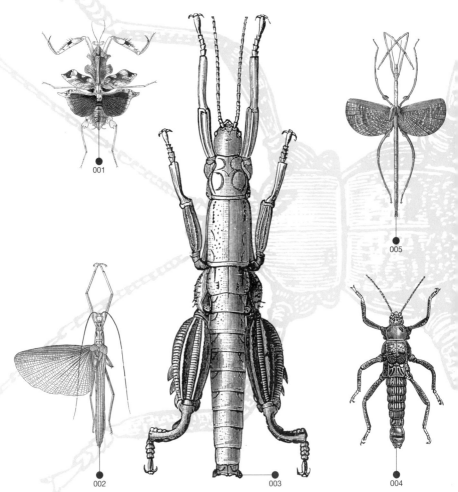

001

002

003

004

005

STICK INSECTS etc

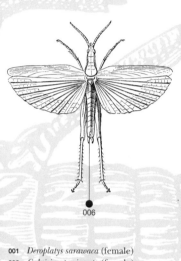

006

The insects here and on the following six pages represent a mixture of mantids, cockroaches, stick insects, and grasshoppers, and perhaps give some indication as to why early taxonomists often mixed them up and placed them all in the Orthoptera. It was the combined work of anatomists, physiologists, paleontologists, and behavioral entomologists of the late 19th and early 20th centuries that finally solved most of the riddles. Today's entomologists are using molecular-biology techniques to unravel the secrets held in their genes and thus explain some of the remaining controversies. Most of the insects on this page are stick insects, but one is a grasshopper and another a bush cricket. Can you spot the difference?

001 *Deroplatys sarawaca* (female)
002 *Calvisia atrosignata* (female)
003 *Eurycantha (Karabidion) australis* (male)
004 *Anisomorpha pardalina*
005 *Palophus centaurus*
006 Grasshopper—*Pyrgomorpha grylloides* ⚊
007 Bush cricket—*Phasmodes ranatriformis* ⚊

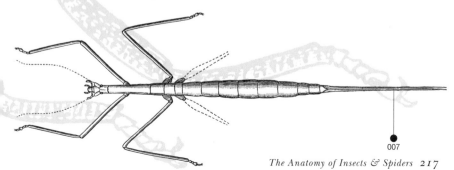

007

The Anatomy of Insects & Spiders **217**

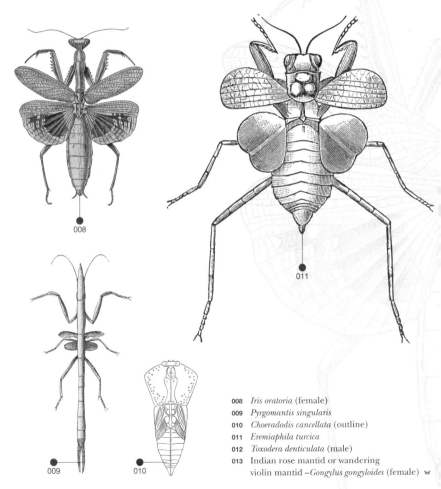

008 *Iris oratoria* (female)

009 *Pyrgomantis singularis*

010 *Choeradodis cancellata* (outline)

011 *Eremiaphila turcica*

012 *Toxodera denticulata* (male)

013 Indian rose mantid or wandering
     violin mantid –*Gongylus gongyloides* (female) ✍

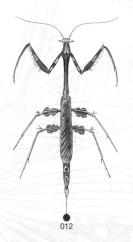

012

The mantids have been important mystic animals in many cultures and civilizations. In his 1634 treatise, *Insectorum Theatrum*, Thomas Mouffet says of them: "They are called Mantes; that is, fortune tellers, either because by their coming (for they first of all appear) they do show the spring to be at hand, as Anacreon, the poet, sang; or else they foretell death and famine, as Caelias, the scholiast of Theocritus writes; or lastly, because it always holds up its fore-feet, like hands praying... So divine a creature is this esteemed, that if a childe aske the way to such a place, she will stretch out one of her feet and show him the right way..."

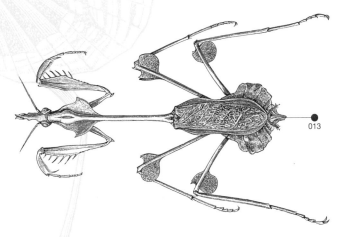

013

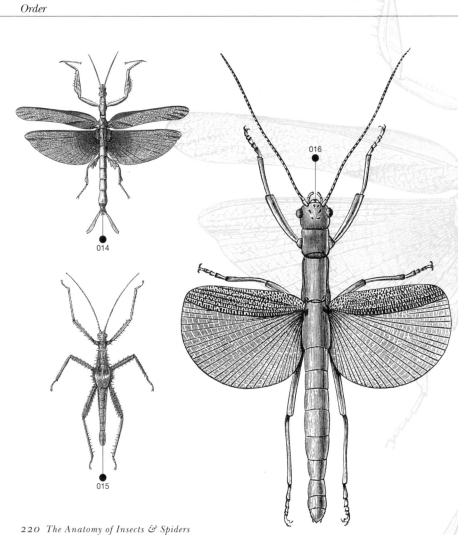

In 1865 the American entomologist Frank Cowan quoted Dr Shaw on the mantids: "Imagination itself can hardly conceive shapes more strange than those exhibited by some species." Many mantids so closely resemble the plants on which they hunt that their names (rose-leaf mantis or pink-flower mantis) were given in recognition of this fact. However, they have proved efficient hunters, even without camouflage. The entomologist C. Denny, writing in 1867, tells how citizens of Melbourne, Australia, placed mantids on their window blinds "so that the rooms may be cleared from flies by the indefatigable [sic] voracity of the Mantis."

014   *Stenophylla cornigera*
015   *Heteropteryx grayi* (male)
016   *Aschipasma catadromus*
017   *Cyphocrania aestuans*
018   *Ceroys saevissima*

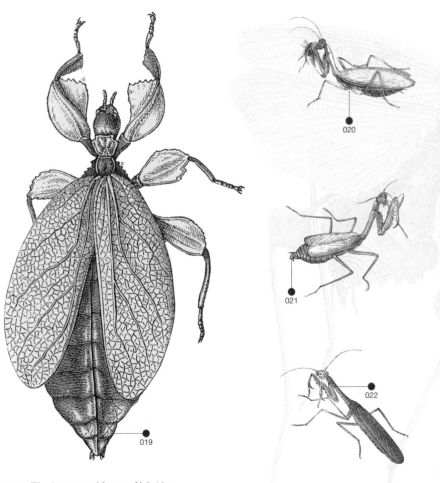

019

020

021

022

Praying mantises are so called because, in their hunting pose, their large, spiny front legs held folded in front of them resemble a person at prayer. A more accurate name might be "preying mantis," because all 2,000 or so species are voracious predators, using their front legs with great speed to catch flies, bees, wasps, butterflies, grasshoppers, and other insects. Male mantises are smaller than the females and often end up as the next meal for them. Many tropical species have bizarre outgrowths on their limbs and bodies, which bear a striking resemblance to the plants on which they hunt. To the Kalahari Bushmen, the mantis is sacred, representing a god of creation.

019 Moving leaf—*Phyllium scythe* (male)
020 Species of mantid
021 Species of mantid
022 Species of mantid
023 Walking leaf—*Phyllium scythe* (male)
024 Species of mantid
025 Common cockroach—*Blatta orientalis* (female)
026 German cockroach—*Blatta germanica*

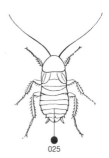

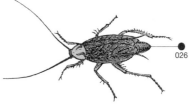

BUGS *Hemiptera*

Hemiptera is the largest of the insect orders in which wings develop externally (exopterygotes), with estimates of described species ranging from 50,000 to 80,000. Their ancestors first appeared in the fossil record before the earliest dinosaurs in the Permian period (about 280 million years ago) and today bugs are found throughout the world. This is an order of stark contrasts and includes the only group of species to truly conquer the open ocean, while others are found in the driest desert environments.

Bugs have had a profound influence on our culture, providing us with waxes, dyes, toys, music, food, crop protection, medicine, fabulous jewelry, a rich folklore, and mystic significance. And yet, year on year, they have also attacked our crops, made our beds a nightmare to sleep in, killed our livestock, and transmitted killer diseases to millions. Some are masters of camouflage; others flaunt their size, voice, and colors with amazing flamboyance. They include some of the largest insects (4¼ inches/11 cm long) and species that may struggle to reach ¹/₂₅ inch/1 mm in length. The one characteristic that unites them are mouthparts that have evolved into a proboscis containing threadlike tubes (stylets), used for piercing and sucking and for deeply penetrating the plants or animals on which they feed.

By the end of the 19th century the American entomologist John B. Smith had published in the journal *Science* an eloquent series of drawings of the structure of the mouthparts. Meanwhile his pioneering contemporaries were bravely exploring the deepest jungles of the Old and the New Worlds, discovering many new species and encountering a barrage of folklore. Some insects from this group were labeled as deadly killers, and, while this turned out to be a myth for many, the threat from other bugs is only too real.

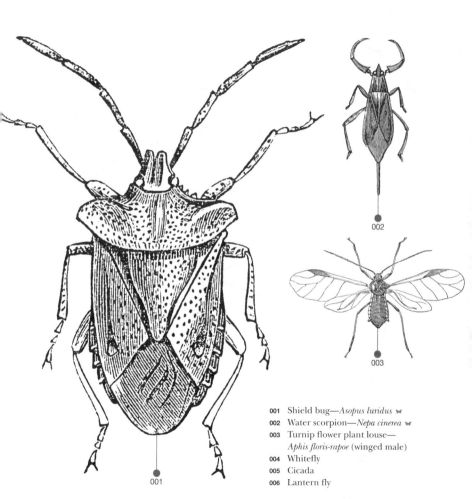

001

002

003

001 Shield bug—*Asopus luridus* ⚭
002 Water scorpion—*Nepa cinerea* ⚭
003 Turnip flower plant louse—
    Aphis floris-rapoe (winged male)
004 Whitefly
005 Cicada
006 Lantern fly

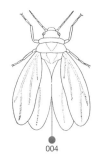

004

The name of the order is derived from the Greek *hemi-*, meaning "half," and *ptera*, meaning "wings." This is because many species have forewings in which the front half differs from the rear. Traditionally, they are divided into two sub-orders, the Heteroptera (in which the front half of the forewing is leathery, the rear half membranous) and the Homoptera (in which the forewings have a membranous texture throughout). The Heteroptera, known as the "true bugs," are a diverse group, adapted to a wide range of environments. They are either plant feeders, predators, or blood-sucking parasites. The Homoptera feed exclusively on plant sap and include among their number familiar groups such as the aphids, whitefly, scale insects, cicadas, froghoppers, leafhoppers, psyllids, and lantern flies.

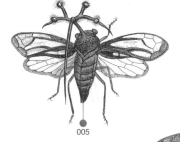

005

006

"Although I am an insect very small, yet with great virtue and endow'd withal."
EDWARD TOPSEL

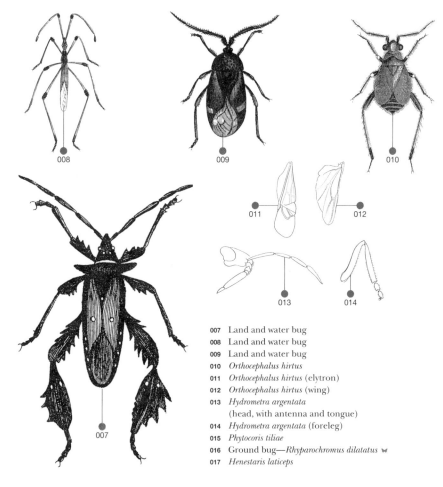

008

009

010

011   012

013   014

007 Land and water bug
008 Land and water bug
009 Land and water bug
010 *Orthocephalus hirtus*
011 *Orthocephalus hirtus* (elytron)
012 *Orthocephalus hirtus* (wing)
013 *Hydrometra argentata*
     (head, with antenna and tongue)
014 *Hydrometra argentata* (foreleg)
015 *Phytocoris tiliae*
016 Ground bug—*Rhyparochromus dilatatus* ✎
017 *Henestaris laticeps*

007

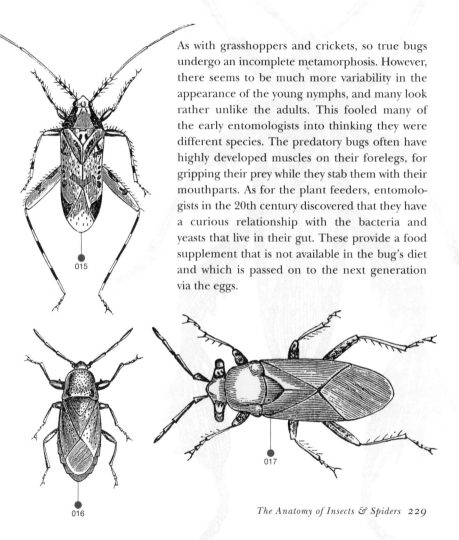

As with grasshoppers and crickets, so true bugs undergo an incomplete metamorphosis. However, there seems to be much more variability in the appearance of the young nymphs, and many look rather unlike the adults. This fooled many of the early entomologists into thinking they were different species. The predatory bugs often have highly developed muscles on their forelegs, for gripping their prey while they stab them with their mouthparts. As for the plant feeders, entomologists in the 20th century discovered that they have a curious relationship with the bacteria and yeasts that live in their gut. These provide a food supplement that is not available in the bug's diet and which is passed on to the next generation via the eggs.

015

016

017

Class	INSECTA
Sub-Class	PTERYGOTA
Division	EXOPTERYGOTA
Order	

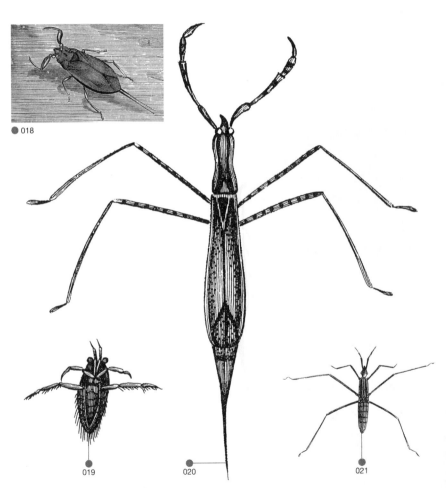

018

019

020

021

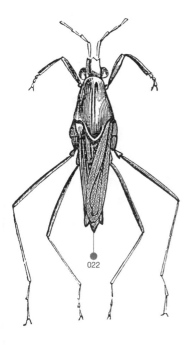

022

Most true bugs live on dry land, but a significant number of them have adopted an aquatic lifestyle, living either on or beneath the surface. The fearsome water scorpion (*no. 018*) has powerful forelegs that resemble scorpion's claws and allow it to seize its unfortunate victim before sucking out its body contents. It breathes at the surface through a long tube attached to its abdomen. The water boatmen (*no. 019*) have long rear legs, which scull them through the water like oars. One group of bugs swim on their backs, coming to the surface to breathe and feed on small fish, tadpoles, or insects and other invertebrates. Another group swim on their fronts, carry air on their abdomens, and feed on microscopic algae.

018 Water scorpion—*Nepa cinerea* ❧
019 Water boatman ❧
020 Water stick insect—*Ranatra linearis* ❧
021 Pond skater ❧
022 *Hydrometra gibbifera*
023 Lesser water boatman—*Corixa geoffroyi* ❧

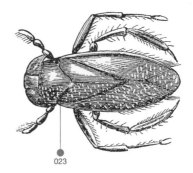

023

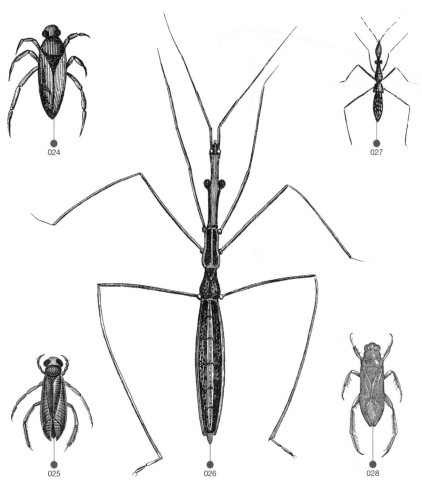

024

027

025

026

028

The great explorer Baron Alexander von Humboldt, on his visit to Mexico ca. 1800, recorded large numbers of eggs being collected on the surface of lakes and sold in the markets as caviar, known as *Axayacat.* These were identified in 1831 by Thomas Say and M. Guerin Meneville as being from three species of water boatmen. The Mexicans farmed the eggs by cultivating a type of sedge upon which the insects readily laid their eggs, floating it in bundles on the lake surface. The water striders or pond skaters live on the surface of the water, feeling for vibrations from insects stranded in the surface film and racing to feed on them, once detected. Some of these live on the open oceans, far from any land.

029

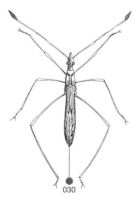

030

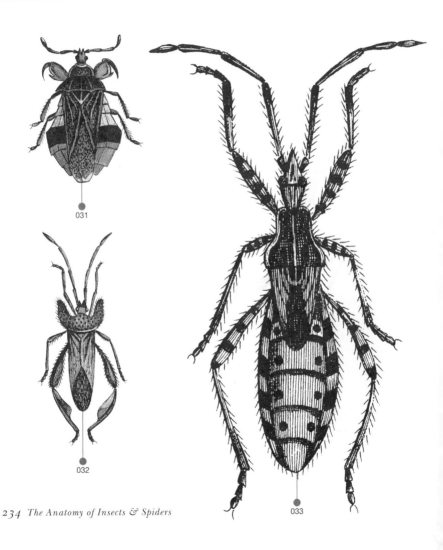

031

032

033

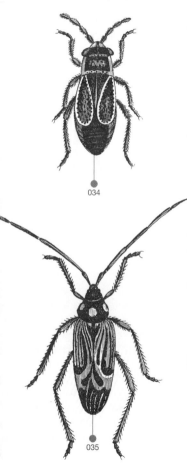

034

Among the land-dwelling species of Heteroptera are some of the most feared of insects, as well as some of the most beneficial. Many are voracious predators of crop pests, and some species are mass-reared to be used as natural pest controllers. Others are themselves crop pests. Kissing bugs, so called because "of a much publicized incident of a young lady who was 'bitten' on the lip…," carry the deadly American form of sleeping sickness known as Chagas's disease. It was probably this disease that led to Darwin's life of ill health. Kissing bugs were greatly feared by 19th-century entomologists, and even today there is no effective cure. They kill 5,000 people each year in Brazil alone and chronically affect millions more.

031–036 Various unidentified land and water bugs ⌦

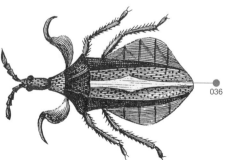

036

035

Class	INSECTA		
Sub-Class		PTERYGOTA	
Division			EXOPTERYGOTA
Order			

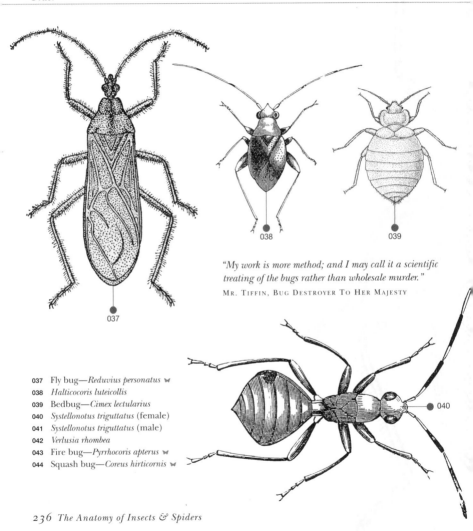

"My work is more method; and I may call it a scientific
treating of the bugs rather than wholesale murder."

MR. TIFFIN, BUG DESTROYER TO HER MAJESTY

037   Fly bug—*Reduvius personatus*
038   *Halticocoris luteicollis*
039   Bedbug—*Cimex lectularius*
040   *Systellonotus triguttatus* (female)
041   *Systellonotus triguttatus* (male)
042   *Verlusia rhombea*
043   Fire bug—*Pyrrhocoris apterus*
044   Squash bug—*Coreus hirticornis*

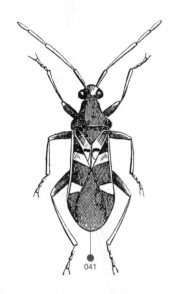

041

Bedbugs (*no. 039*) are blood-sucking parasites of mammals and birds. Two species affect humans, spending the day hiding in crevices in bedrooms. Remarkably, male bedbugs do not deposit their sperm in the female's reproductive tract, but puncture her abdomen and inject it into the body cavity. They were said to have been introduced into Britain on imported lumber used after the Fire of London in 1666, and to be "particularly prevalent in dirty houses." However, there is a remarkable account dating from 1865 by a Mr. Tiffin of Messrs Tiffin & Son, Bug Destroyers to Her Majesty, in which he claims, "I was once at work on the Princess Charlotte's own bedstead"; he found a single bug, which "I think…looked all the better for having tasted royal blood."

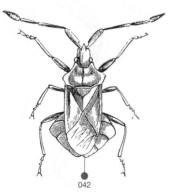

042

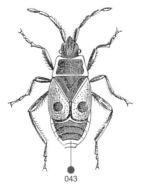

043

044

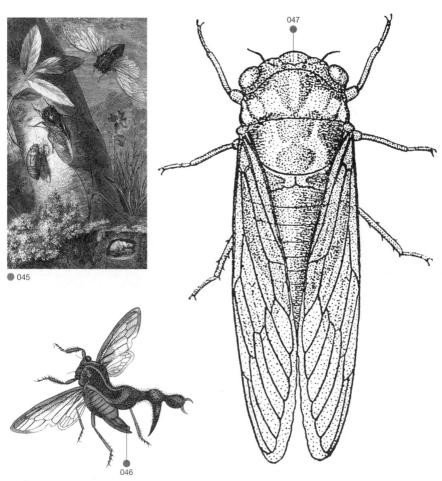

045

046

047

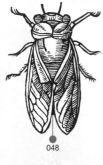

048

049

Cicadas are from the Homoptera. Both the ancient Greeks (who knew them as *Tettix*) and Chinese kept them in cages to enjoy the males' song. The Chinese regarded them as symbols of immortality and Greek children used them as toys. In his 1865 book Frank Cowan states: "*Tettix* seem to have been the favorites of every Greek bard, from Homer and Hesiod to Anacreon and Theocritus....But the old witticism, attributed to the incorrigible...Xenarchus, gives quite a different reason to account for the supposed happiness of these insects: Happy the Cicadas' lives, Since they all have voiceless wives."

045–052 Cicadas—Cicadidae

050

051

052

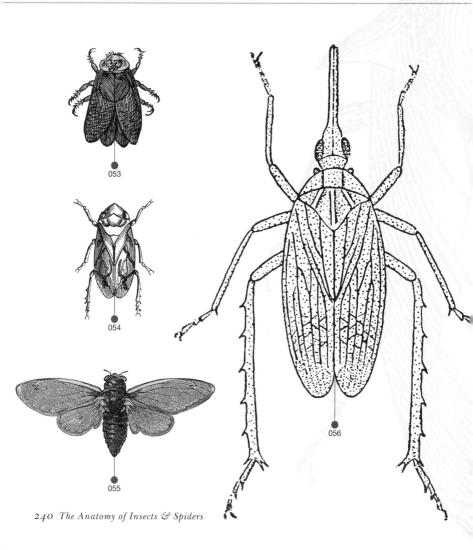

053

054

055

056

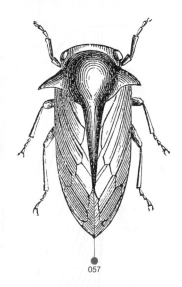

057

One of the biggest entomological myths dates from the 18th century and involves the lantern fly, *Fulgora laternaria* (*no. 059*). One story arose from a Mme Merim, who was living in Surinam ca. 1700. Her account tells of natives bringing her lantern flies, which she claims shone by night with a "flame of fire." Another story from Brazil, recounted by John Branner in 1885 in the *American Naturalist*, states: "it has great powers of flight, and when, in its wild career, it strikes any living object—if an animal, no matter how large and powerful—it falls dead upon the spot; if a tree it soon withers and dies." Branner later bravely traveled to Brazil and proved both accounts to be fallacies.

053 Cicada
054 *Ptyelus bifasciatus*
055 Cicada
056 Lantern fly—*Fulgora* sp
057 Eared hopper—*Centrotus cornutus*
058 Spotted hopper—*Cercopis dorsivittata*
059 Peanut-head bug or lantern fly—*Fulgora laternaria*

058

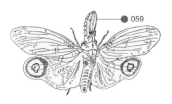

059

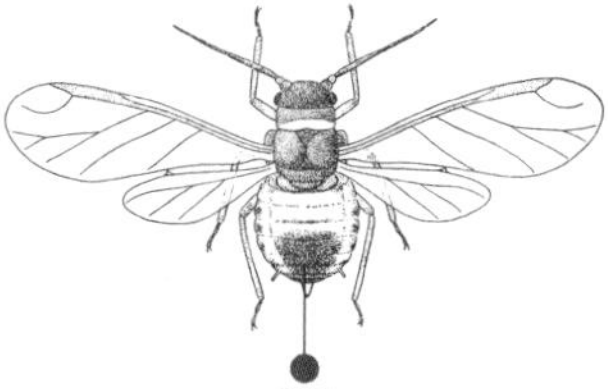

060

060 Winged aphid—Aphididae
061 Currant aphid—*Myzus ribis* (male)
062 Currant aphid—*Myzus ribis* (winged female)
063 Currant aphid—*Myzus ribis* (apterous female)
064 Black bean aphid—*Aphis fabae* (winged male)
065 Black bean aphid—*Aphis fabae* (wingless female)
066 Turnip flower plant louse—*Aphis floris-rapae* (wingless female)
067 American blight—*Schizoneura lanigera* (wingless female)
068 American blight—*Schizoneura lanigera* (winged female)
069 American blight infested twig

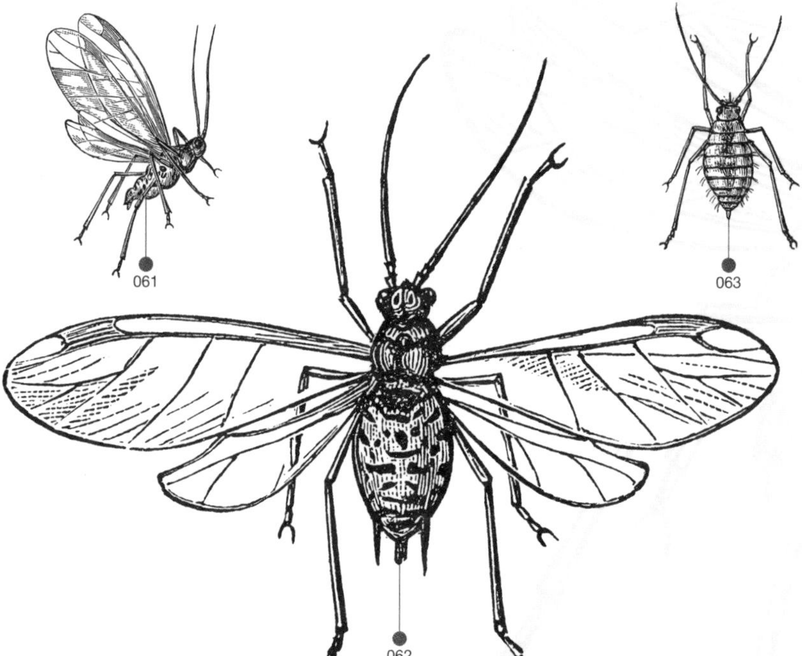

061

063

062

BUGS 9

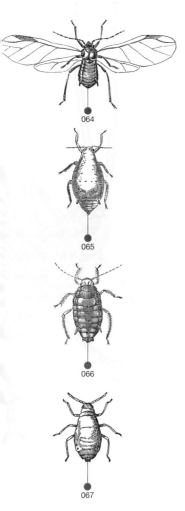

064

065

066

067

Aphids (*no.s 060–065*) have complex life cycles, overwintering as eggs on woody plants, while spending the summer on vegetables and flowers. They have winged and non-winged generations, and a sexual generation and many non-sexual generations each year. The wingless forms of the latter are phenomenal breeding machines, producing their first offspring only a few days after birth, and thereafter bearing a live nymph every hour or so. Each live youngster is effectively a grandmother at birth, bearing a nymph inside her ovaries, which in turn has an egg inside it. One 20th-century entomologist worked out that one aphid could give rise to 600 billion offspring in a single summer— equivalent to the weight of 10,000 humans!

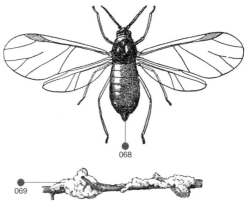

068

069

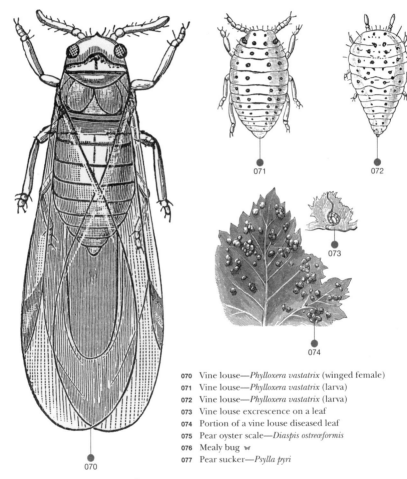

070 Vine louse—*Phylloxera vastatrix* (winged female)
071 Vine louse—*Phylloxera vastatrix* (larva)
072 Vine louse—*Phylloxera vastatrix* (larva)
073 Vine louse excrescence on a leaf
074 Portion of a vine louse diseased leaf
075 Pear oyster scale—*Diaspis ostreæformis*
076 Mealy bug ✶
077 Pear sucker—*Psylla pyri*

BUGS *10*

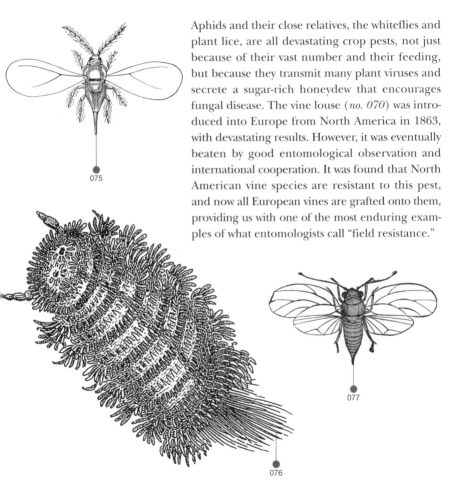

Aphids and their close relatives, the whiteflies and plant lice, are all devastating crop pests, not just because of their vast number and their feeding, but because they transmit many plant viruses and secrete a sugar-rich honeydew that encourages fungal disease. The vine louse (*no. 070*) was introduced into Europe from North America in 1863, with devastating results. However, it was eventually beaten by good entomological observation and international cooperation. It was found that North American vine species are resistant to this pest, and now all European vines are grafted onto them, providing us with one of the most enduring examples of what entomologists call "field resistance."

075

077

076

Class	INSECTA		
Sub-Class		PTERYGOTA	
Division			EXOPTERYGOTA
Order			

078 Cottony cushion scale—
    Icerya purchasi (male) ⛥
079 Mealy bug—
    Pseudococcus sp (female)
080 Pear psyllid—*Psylla pyricola*
    (nymph in last instar)
081 Apple psyllid—*Psylla mali*
082 Greenhouse whitefly—
    Trialeurodes vaporariorum (imago)
083 Greenhouse whitefly—*Trialeurodes
    vaporariorum* (larva in first instar)
084 Greenhouse whitefly caudal
    breathing fold

Scale insects have provided us with dyes such as cochineal and waxes such as shellac. Even the biblical "manna" is thought to have been honeydew from these insects. In 1888 the Australian cottony cushion scale (*no. 078*) was devastating the citrus crops in California. An entomologist called Albert Koebele from the U.S. Department of Agriculture (USDA) was dispatched to Australia to search for natural enemies of this pest. He returned in 1889 with some vedalia beetles (a type of ladybug) and introduced them into the citrus groves. Within months the beetles had the pest under control and have kept it so to this day. This was hailed as being the first great biological-control success of modern times and has since inspired many similar experiments.

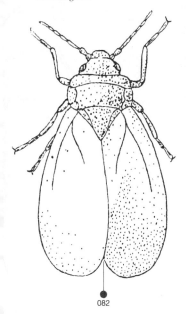

082

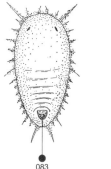

083

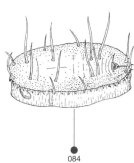

084

SPIDERS *Araneae*

Spiders are often misclassified under the label of "insect," and indeed they are in the same phylum as insects, but they are in the Arachnida, a class they share with scorpions and mites. The implication is that spiders originated in the Devonian period, about 100 million years before the appearance of flies. *Attercop*, we are told by W. S. Bristowe in *The World of Spiders* (1958), is an Anglo-Saxon name for spider and means "poison-head," while *damhan allaidh*, a Gaelic name for these eight-legged creatures, means "little wild deer." These skillful individuals, often living in close association with humans (for instance, house spiders), also inhabit ants' nests. And aquatic habitats in Asia and Europe have been invaded by the so-called water spiders—hunting spiders with "diving bells" full of air.

Spiders often gain a bad press, and, while the more beautiful creatures (some may argue) represented in this book, such as butterflies and ladybugs, are often employed as positive images in advertising, spiders are more often than not associated with fear and loathing. Yet one of the most beautiful sights on a dewy morning is a spider's web dripping with pearl-like drops that glisten in the early morning rays. These magnificently woven insect-catchers have been implicated in tales of protection, such as the one on the right, retold by Frank Cowan. Other web stories indicate medical uses for the fine, silken strands. Bristowe tells us that spiders' webs were used to impede blood flow as early as Dioscorides in the first century AD—a technique also reportedly utilized in Britain.

"*It is related in the life of Mohammed, that when he and Abubeeker were fleeing for their lives before Coreishites, they hid themselves for three days in a cave, over the mouth of which a Spider spread its web, and a pigeon laid two eggs there, the sight of which made the pursuers not go in search for them.*"

FRANK COWAN, 1865

ANATOMY OF A SPIDER

The body of a spider is made up of 2 main regions; the abdomen or opisthosoma and the head region or cephalothorax. Spiders have 8 legs projecting from the cephalothorax or prosoma, which itself is protected by a hardened carapace. The spinnerets or spinners are situated at the posterior end of the abdomen and are the organs through which spider-silk is emitted.

Spiders may have 2, 4 or 6 eyes but most species have 8. Differences in eyesight correspond to the differences in lifestyles. Hunting spiders, for example, have well-developed eyesight over short distances, but in contrast, some species of cave-dwelling spiders have no eyesight at all.

In common with insects, spiders have an open blood system, which means that blood flows in between the organs, in the open spaces. Spiders' blood is not only important for transporting oxygen and nutrients, and so on, but also for molting, a process known as "ecdysis" where the outer skin is shed. Tarantulas may live up to 25 years and for them molting takes place yearly.

001 Palp
002 Eyes
003 Chelicera

LEG SEGMENTS
004 Coxa
005 Trochanter
006 Femur
007 Patella
008 Tibia
009 Metatarsus
010 Tarsus
011 Tarsal claws

012 Abdomen
013 Spinnerets
014 Anal tubercle
015 Fovae
016 Cephalothorax

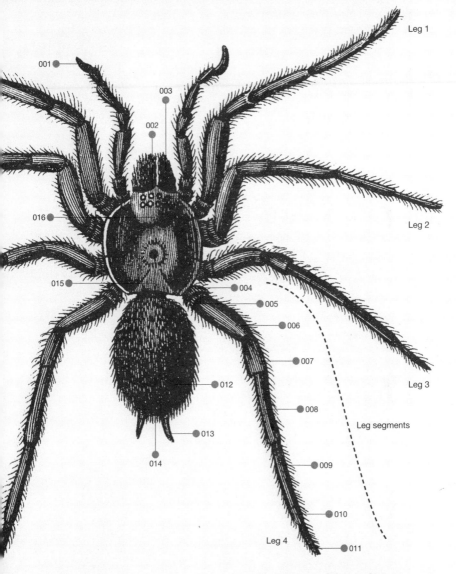

001

003

002

Leg 1

Leg 2

016

015

004

005

006

007

Leg 3

012

008

Leg segments

013

009

014

010

Leg 4

011

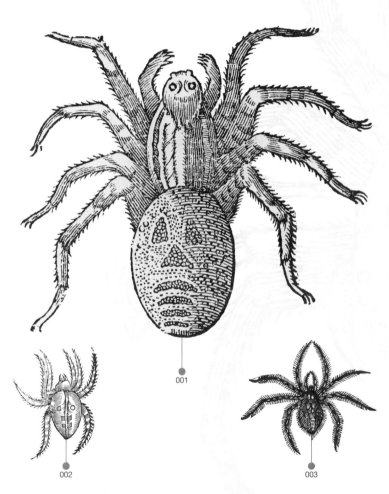

001

002

003

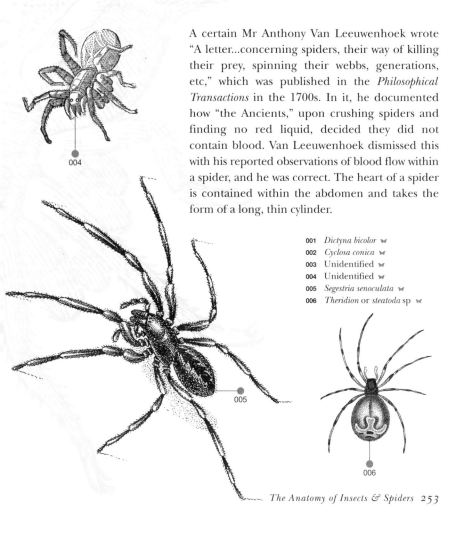

A certain Mr Anthony Van Leeuwenhoek wrote "A letter...concerning spiders, their way of killing their prey, spinning their webbs, generations, etc," which was published in the *Philosophical Transactions* in the 1700s. In it, he documented how "the Ancients," upon crushing spiders and finding no red liquid, decided they did not contain blood. Van Leeuwenhoek dismissed this with his reported observations of blood flow within a spider, and he was correct. The heart of a spider is contained within the abdomen and takes the form of a long, thin cylinder.

001 *Dictyna bicolor*
002 *Cyclosa conica*
003 Unidentified
004 Unidentified
005 *Segestria senoculata*
006 *Theridion* or *steatoda* sp

010

Thomas Mouffet included spiders in his *Theater of Insects* and placed them in a simple wingless insect classification, which was divided into two groups: those frequenting water and those of the earth. The earth group was itself divided into "some with feet" and "some without feet," and spiders came under the label of those that "goe with eight feet, scorpions and spiders." Interestingly, W. S. Bristowe surmises that the "Miss Muffet" of the beloved nursery rhyme (concerning the said Miss, a tuffet, and a spider) is none other than Thomas Mouffet's (Muffet's) own daughter, Patience.

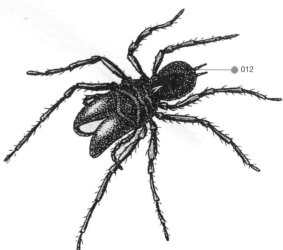

012

011

007 Unidentified 🕷
008 *Sparassus smaragdulus*
009 Tarantula—*Stichoplastus* sp 🕷
010 Unidentified 🕷
011 *Atypus affinis* (female) 🕷
012 *Atypus affinis* (male) 🕷

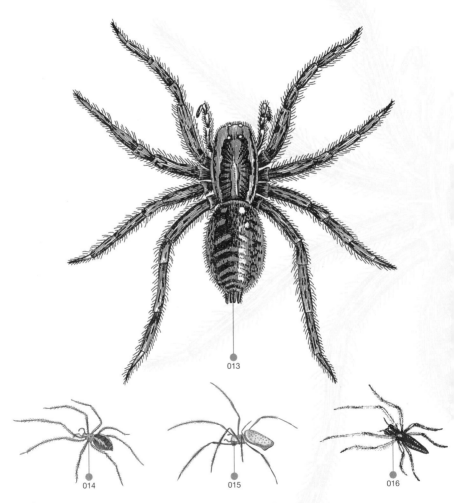

013

014

015

016

017

"It is unfortunate that spiders should have a prominent long-legged ambassador who creates prejudice against his race by invading the privacy of our bathrooms."
W. S. BRISTOWE

013 Unidentified
014 *Linyphia triangularis*
015 *Nephila* sp
016 *Tetragnatha extensa*
017 Unidentified
018 Unidentified

Gossamer, or spiders' webbing or "sun-dew webs" (as quoted by Frank Cowan in 1865 from a Scottish dictionary) have a tendency to float in still air and are said to have been considered as a form of chemical warfare around the time of the two world wars. Earlier than this, in the 17th century, clouds were thought to be formed from the same material. Pliny explains: "in the year that L. Paulus and C. Marcellus were consuls it rained wool…," in what is said by W. S. Bristowe to be the "earliest known reference to the gossamer flecks." Aboard the *Beagle*, Charles Darwin detailed his observations on this subject, writing, "On several occasions, when the *Beagle* has been within the mouth of the Plata, the rigging has been coated with the web of the Gossamer Spider."

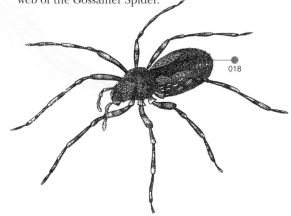

018

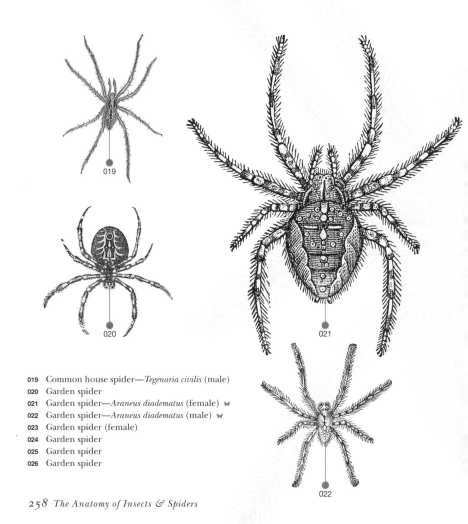

019 Common house spider—*Tegenaria civilis* (male)
020 Garden spider
021 Garden spider—*Araneus diadematus* (female) ⤳
022 Garden spider—*Araneus diadematus* (male) ⤳
023 Garden spider (female)
024 Garden spider
025 Garden spider
026 Garden spider

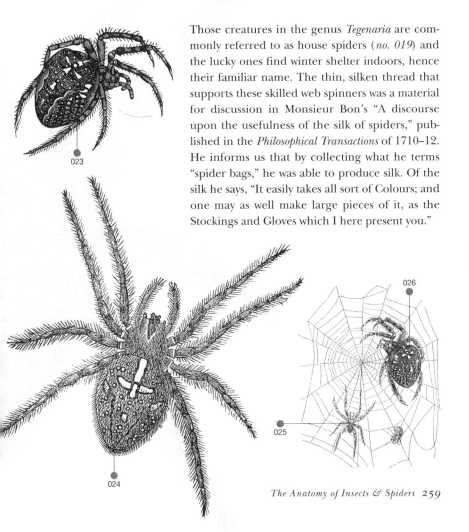

Those creatures in the genus *Tegenaria* are commonly referred to as house spiders (*no. 019*) and the lucky ones find winter shelter indoors, hence their familiar name. The thin, silken thread that supports these skilled web spinners was a material for discussion in Monsieur Bon's "A discourse upon the usefulness of the silk of spiders," published in the *Philosophical Transactions* of 1710–12. He informs us that by collecting what he terms "spider bags," he was able to produce silk. Of the silk he says, "It easily takes all sort of Colours; and one may as well make large pieces of it, as the Stockings and Gloves which I here present you."

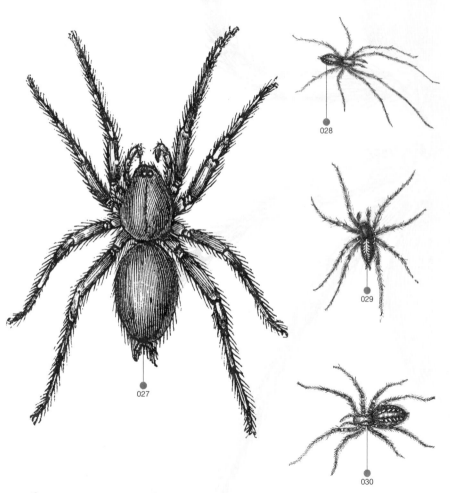

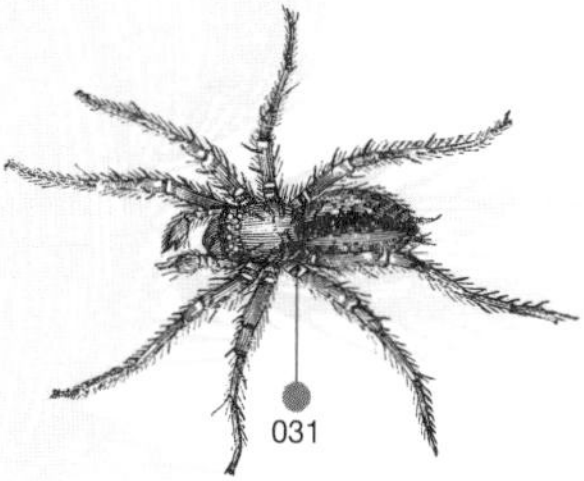

031

W. S. Bristowe tells us that one species of long-legged house spider earned the popular name of cardinal spider (a relation of the *Tegenaria* species pictured here) because of a petrified Cardinal Wolsey, who took a great dislike to these hairy residents of Hampton Court in the early 16th century. He was not alone in his aversion—long legs, hair, and skittish movements apparently being the main reasons in this respect. The fact that some spiders administer painful (and even sometimes fatal) bites hinders their acceptance. The bite of the Mediterranean black widow spider is exceptionally painful, but rarely fatal.

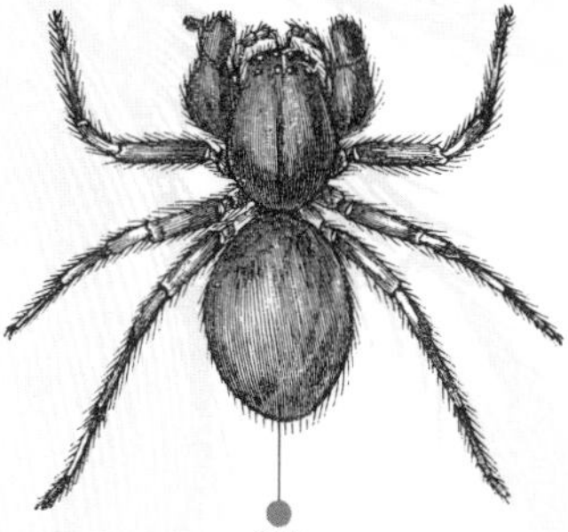

032

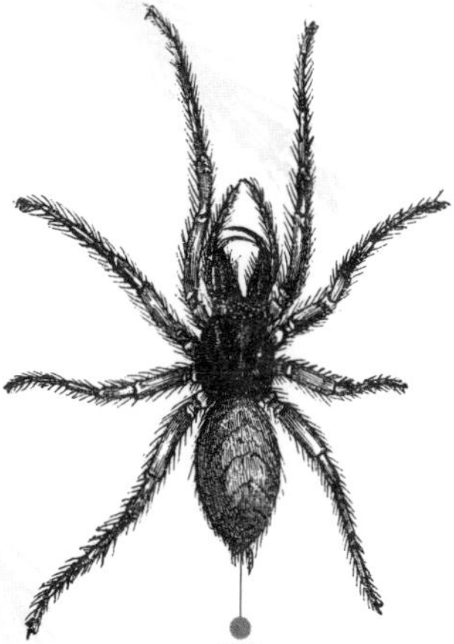

033

027 *Nops guanabocoae*
028 House spider—
Tegenaria domestica (male)
029 *Agalena labyrinthica*
030 House spider—
Tegenaria domestica (female)
031 *Coelotes saxatilis*
032 *Otiothops walckenaeri*
033 *Dysdera erythrina*

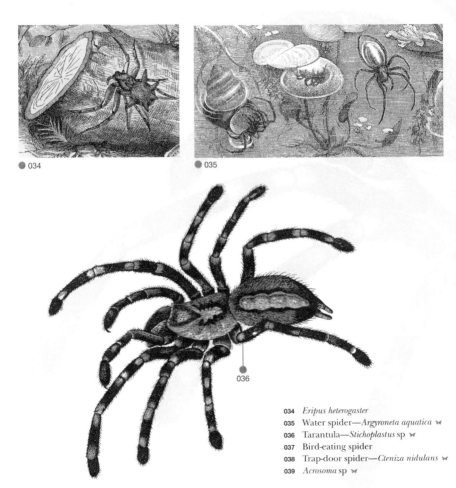

034

035

036

034 *Eripus heterogaster*
035 Water spider—*Argyroneta aquatica* ✤
036 Tarantula—*Stichoplastus* sp ✤
037 Bird-eating spider
038 Trap-door spider—*Cteniza nidulans* ✤
039 *Acrosoma* sp ✤

037

Most species of bird-eating spiders (*no. 037*) feed on insects, but a name such as the "bird-eating spider" is full of foreboding and seems quite appropriate for the colossal size of these hunters. With an impressive legspan and a weight of about 3 ounces/85 g, it is little wonder that these tropical and sub-tropical inhabitants are given the name "baboon spiders" in Africa. The largest spider in the world is the South American Goliath bird-eater, with an impressive legspan just shorter than twice the height of this book. A large, hairy spider often invokes the name "tarantula" (*no. 036*) and indeed these particular beasts are commonly called by this term, however, Michael Chinery informs us in *Spiders* that the name is correctly assigned to the European wolf spider.

038

039

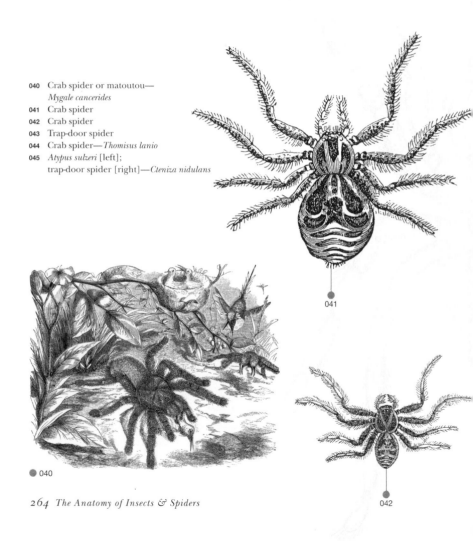

041

040

042

043

W. S. Bristowe introduces us to the theory that music is "enjoyed" by spiders, by explaining that the vibrations that prey make on a spider's web may be reproduced with the aid of a tuning fork or violin. Such historical anecdotes as the following one, reproduced in Frank Cowan's 1865 book, are extremely interesting: A prisoner in the Bastille named Pelisson fed flies to a spider, which emerged to eat the food provided when the prisoner's companion played the bagpipes. Interestingly, Robert E. C. Stearns, in a paper written for the *American Naturalist* in 1890, entitled "Instances of the Effects of Musical Sounds on Animals," tells of an article that chronicled the activities of an Englishman in New York who liked to discipline creatures—including spiders—using music.

045

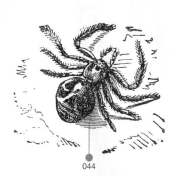

044

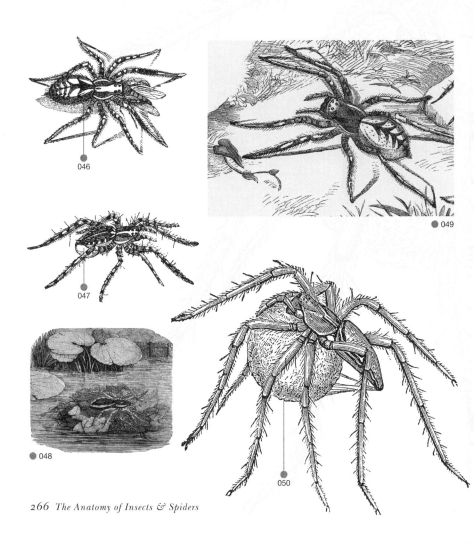

046

049

047

048

050

● 051

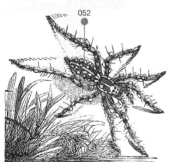

052

Maternal wolf spiders are ferociously protective of their egg sacs (*no. 050*) and carry their hatched babies for about a week. Charles Darwin wrote of these spiders during his acclaimed voyage aboard the *Beagle* in the 1830s. And the terms *taranto, tarantella,* and *tarantism* are used in an interesting anecdote chronicled by Michael Chinery concerning tarantulas and dancing. Taranto is an Italian town associated with the tarantula, and the story goes that, in order to relieve oneself of a painful bite administered by these spiders, one must dance until one drops; *tarantella* was the name given to this dance and the later addition of music in the Middle Ages; and tarantism was the "mass hysteria" that ensued when spectators joined in.

046 Wolf spider—*Lycosa andrenivora*
047 Wolf spider—*Lycosa saccata* (female)
048 Raft spider—*Dolomedes fimbriatus* ⋈
049 European wolf spider—*Lycosa tarantula*
050 Wolf spider with egg sac—*Dolomedes* sp (female)
051 Raft spider—*Dolomedes fimbriatus* ⋈
052 Wolf spider—*Lycosa saccata* (male)

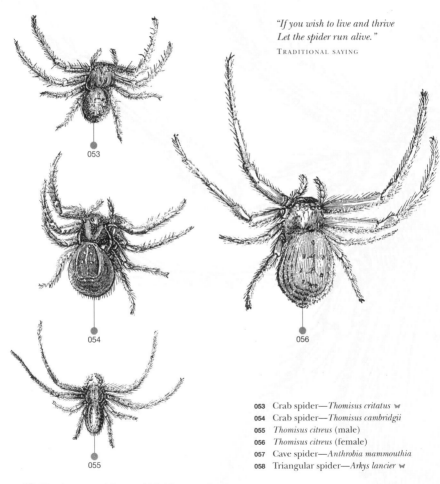

*"If you wish to live and thrive
Let the spider run alive."*

TRADITIONAL SAYING

053 Crab spider—*Thomisus critatus* 🦋
054 Crab spider—*Thomisus cambridgii*
055 *Thomisus citreus* (male)
056 *Thomisus citreus* (female)
057 Cave spider—*Anthrobia mammouthia*
058 Triangular spider—*Arkys lancier* 🦋

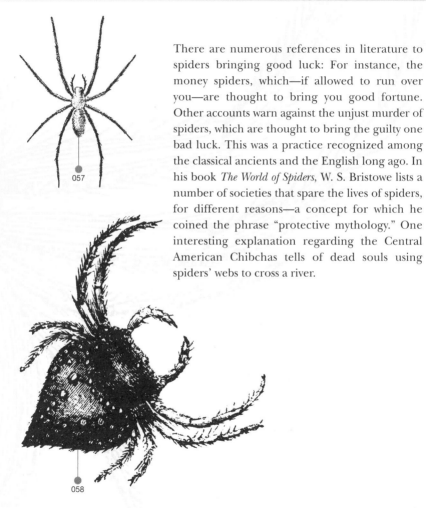

057

058

There are numerous references in literature to spiders bringing good luck: For instance, the money spiders, which—if allowed to run over you—are thought to bring you good fortune. Other accounts warn against the unjust murder of spiders, which are thought to bring the guilty one bad luck. This was a practice recognized among the classical ancients and the English long ago. In his book *The World of Spiders*, W. S. Bristowe lists a number of societies that spare the lives of spiders, for different reasons—a concept for which he coined the phrase "protective mythology." One interesting explanation regarding the Central American Chibchas tells of dead souls using spiders' webs to cross a river.

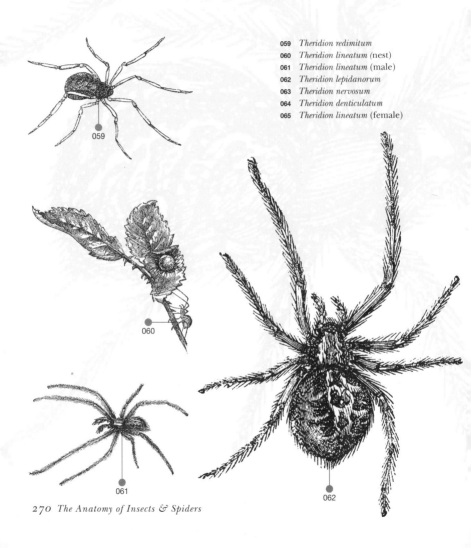

059 *Theridion redimitum*
060 *Theridion lineatum* (nest)
061 *Theridion lineatum* (male)
062 *Theridion lepidanorum*
063 *Theridion nervosum*
064 *Theridion denticulatum*
065 *Theridion lineatum* (female)

059

060

061

062

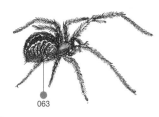

063

064

In this book we have been reminded that insects were often regarded as useful medical remedies throughout history. Here we see that spiders were, too. Indeed, W. S. Bristowe documents the belief that, in Dioscorides's and Pliny's day, spiders would chase away disease if they were worn around the neck. Frank Cowan in 1865 relates: "A third kind of Spiders, says Pliny, also known as the "phalangium," is a spider with a hairy body, and a head of enormous size. When opened, there are found in it two small worms, they say: these attached in a piece of deer's skin, before sunrise, to a woman's body, will prevent conception…" In fact *Phalangium* is a genus name of the harvestmen, which are only distantly related to spiders.

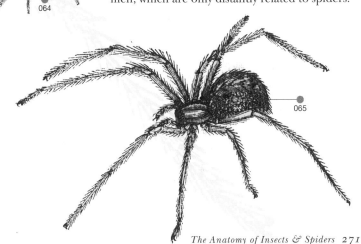

065

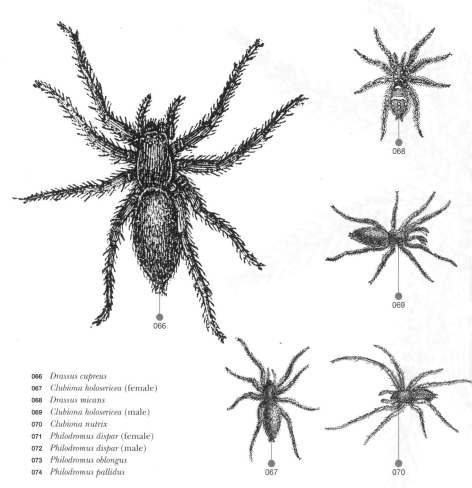

066 *Drassus cupreus*
067 *Clubiona holosericea* (female)
068 *Drassus micans*
069 *Clubiona holosericea* (male)
070 *Clubiona nutrix*
071 *Philodromus dispar* (female)
072 *Philodromus dispar* (male)
073 *Philodromus oblongus*
074 *Philodromus pallidus*

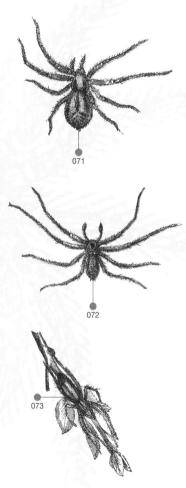

071

072

073

Undoubtedly, eating spiders has little appeal, but from the 16th century onwards the ingestion of spiders was a popular practice among fever sufferers. Frank Cowan tells us how the New Caledonian populace, Native Americans, and Aboriginal Australians ate spiders. Having said this, he also relates the story of a group of Florentine monks who drank wine from a container that had a drowning spider in it, and died. "The people of Maniana, south of Gambia and Senegal, are cannibals. They eat Spiders, Beetles, and old men," says Molien, again according to the American entomologist.

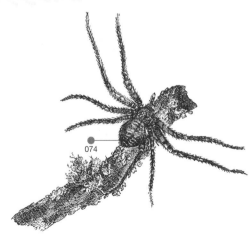

074

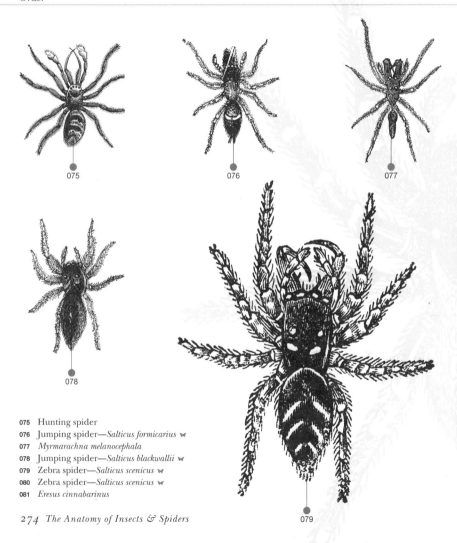

075　Hunting spider
076　Jumping spider—*Salticus formicarius* ⚭
077　*Myrmarachna melanocephala*
078　Jumping spider—*Salticus blackwallii* ⚭
079　Zebra spider—*Salticus scenicus* ⚭
080　Zebra spider—*Salticus scenicus* ⚭
081　*Eresus cinnabarinus*

*"A little Head and body small
With slender feet and very tall,
Belly great, and from thence came all
The web it spins."*

THOMAS MOUFFET

Rather endearingly, Frank Cowan describes spiders as "living barometers." Indeed—both in the days of Pliny and fairly recently in Europe and America—spiders were widely thought to be prognostic of the weather. W. S. Bristowe eloquently explains in his 1958 account that "Omens of good weather include an orb-weaver in the centre of its web, the appearance of a sheet-builder at the mouth of her retreat, the building of orb-webs immediately after the rain has stopped, and the descent of gossamer on fields or ships at sea." Depicted here are spiders from the Salticidae family, commonly called jumping spiders. The British naturalist Alfred Wallace described them as "more like jewels than spiders," in reference to their often iridescent bodies.

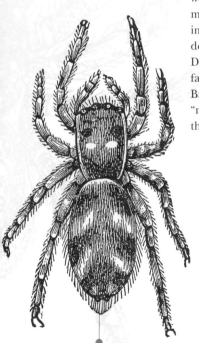

080

081

GLOSSARY *1*

ABDOMEN: the hindmost of the three main body sections of an insect. It contains most of the digestive organs and the reproductive organs. It may also be used to describe the opisthosoma of a spider.

ANATOMIST: one who studies the structure of an organism.

ANTENNA: (plural: antennae) sometimes also called the "feelers." The sensory organs on the head, responsible for smelling, sensing heat, humidity, and wind direction.

BEETLE

APICULTURE: beekeeping.

ASEXUAL: reproduction without copulation between male and female. Many of the insect groups are able to reproduce in this way; in some species males are unknown. *See also* Parthenogenesis.

BATESIAN MIMICRY: where the mimic is a defenseless organism which gains protection by looking similar to a model that is distasteful or that is otherwise protected from predation.

BIOLOGICAL CONTROL: the use of natural enemies and diseases in order to control pests, diseases, or weeds.

BLACK DEATH: an outbreak of the plague during the 14th century (*cf.* Bubonic plague).

BUBONIC PLAGUE: a disease characterized by inflamed swellings and fever. Rats may be carriers of the disease and fleas are able to pass it on to humans.

CAUDAL: of the tail.

CELLULOSE: the chief structural material of cell walls in the plant kingdom.

CEPHALOTHORAX: *see* Prosoma.

CLASS: a category within the Linnaean classification system describing a group of closely related orders. It may be further divided into sub-classes and divisions.

COCOON: a case made of silk that protects the pupae of insects, especially moths. It is constructed by the larva before it pupates.

COMPLETE METAMORPHOSIS: describes a life cycle where the immature stages differ greatly from the adult (e.g. butterflies, wasps, beetles). *See also* Endopterygote, Hemimetabolous.

DEVONIAN PERIOD: a geological period that occurred between about 400 and 350 million years ago.

DIVISION: a category within the Linnaean classification system that, for animals, describes a sub-group of closely related orders.

ELYTRON: (plural: elytra) the hardened front wing of a beetle or earwig.

ENDOPTERYGOTE: an insect in which the wings develop inside the body in the young insect before it turns into a pupa. The immature differs markedly from the adult and is called a larva.

ENTOMOLOGIST: one who studies insects and other closely related invertebrates.

EXOPTERYGOTE: an insect in which the wings develop on the surface of the body. There is no pupal stage in the life cycle and the young are called nymphs.

FAMILY: a category within the Linnaean classification system that describes a group of closely related genera.

FECUNDITY: the number of offspring produced.

GALL: a growth produced by plants in response to an unrelated stimulus. Insects that induce the formation of cells are known as zoocecidia.

GENUS: (plural: genera) a category within the Linnaean classification system that describes a group of closely related species.

GOSSAMER: spiders' webs.

HALTERES: balancing structures of the two-winged or true flies (order Diptera). The hindwings of Diptera are reduced to these.

HEMIMETABOLOUS: having an incomplete life cycle with no pupal stage (e.g. dragonflies, true bugs, and grasshoppers). *See also* Incomplete metamorphosis, Exopterygote.

HOLOCENE PERIOD: a geological period referring to approximately the last 10,000 years.

HOLOMETABOLOUS: having a complete life cycle with a pupal stage. *See also* Complete metamorphosis, Endopterygote.

HONEYDEW: a sugary substance excreted by many sap-feeding insects. It contains approximately 98 percent sucrose.

IMAGO: (plural: imagines) the adult stage of an insect's life cycle.

GRASSHOPPER

INCOMPLETE METAMORPHOSIS: describes insect life cycles where the immature stages may not differ greatly from the adults and do not have a pupal stage (e.g. true bugs, grasshoppers, and dragonflies). *See also* Exopterygote, Hemimetabolous.

INSTAR: the stage in an insect's life cycle between any two molts. A hatchling from an egg is described as a first instar, the adult as the final instar.

GLOSSARY 2

INVERTEBRATE: any animal without a backbone (e.g. insects, spiders, worms, and jellyfish).

KINGDOM: the ultimate category within the Linnaean classification system, describing a group of closely related phyla.

LAMELLA: (plural: lamellae) a thin layer.

LARVA: (plural: larvae) an immature insect that differs markedly in appearance and lifestyle from the adult (*cf.* Nymph), e.g. the caterpillar (larva) of a butterfly (adult).

LEPIDOCHROMY: a practice whereby real butterflies are employed to produce an illustration.

MANDIBLE: the jaw of an insect.

BUTTERFLY

METAMORPHOSIS: the changes that occur during the life cycle of any living thing. *See also* Complete metamorphosis, Incomplete metamorphosis.

MIMICRY: *see* Batesian mimicry.

NEARCTIC: denotes an area of the world that comprises Greenland and North America.

NEO-TROPICAL: of South American origin.

NYMPH: an immature insect that does not change into a pupa before becoming adult. In many species nymphs superficially resemble the adult, but wings are not fully formed. Nymphs normally have the same lifestyle and diet as the adult (*cf.* Larva).

OPISTHOSOMA: the hind-most body section of a spider. It contains most of the digestive organs and the reproductive organs. It may also be called the abdomen.

ORDER: a category within the Linnaean classification system that describes a group of closely related families and/or superfamilies.

OVIPOSITOR: the egg-laying organ of a female insect.

PALEONTOLOGIST: one who studies fossils and extinct species.

PARASITE: an organism living in or upon another organism and deriving nourishment from it, to the detriment of the host.

PARTHENOGENESIS: a type of reproduction where eggs develop into nymphs or larvae without having been fertilized.

PHEROMONE: an odor released by an individual to act as a signal to another individual of the same species. It may represent a sexual attractant (most insects), an alarm signal (e.g. aphids), aggregation signal, and so on.

PHYLUM (plural: phyla): a category within the Linnaean classification system describing a group of closely related classes.

PHYSIOLOGIST: one who studies the functions of organs, tissues, and cells within an organism.

PREPUPA: the resting stage undergone by many larvae before turning into pupae.

PROBOSCIS: mouthparts drawn out into a long tube.

PROLEGS: stumpy legs found on the hind region of the caterpillars of butterflies and sawflies.

PROSOMA: the front section of a spider bearing the head and thorax. It may also be called the cephalothorax.

PROTOZOAN: a single-celled member of the former phylum Protozoa, which were classified as animals but are now part of the kingdom Protista (neither animal, fungus, plant, nor bacterium).

PUPA (plural: pupae): a stage between larva and adult during metamorphosis. The pupa does not feed and does not usually move about. It is during this stage that the body is completely rebuilt into that of the adult.

SERICULTURE: the keeping of silk moths for silk production.

SEXUAL: reproduction requiring fertilization of an egg.

SPECIES: the "lowest" category within the Linnaean classification system, describing a population (or series of populations) of closely related organisms. The "biological species concept" is more narrowly defined as a group of individuals capable of breeding freely with each other, but not with members of other species.

STRIDULATION: production of sound by rubbing two parts of the body together.

STYLET: needle-like piercing mouthparts.

FLEA

SUPERFAMILY: a category within the Linnaean classification system describing a group of closely related families.

TAXON (plural: taxa): a unit within a classification system (e.g. a family or genus).

TAXONOMIST: one who studies the classification of organisms.

TERTIARY PERIOD: a geological period that occurred between approximately 70 million and 1 million years ago.

THORAX: the middle of the three major sections of an insect's body. The legs and wings, if present in the insect, are attached to the thorax.

USDA: United States Department of Agriculture.

BOOKS

BARNES, CALOW & OLIVE, *The Invertebrates*, Blackwell Science, Oxford, 1993

BRISTOWE, W. S., *The World of Spiders*, William Collins, London, 1958

CARTER, D., *Butterflies and Moths*, Dorling Kindersley, London, 2000

CHINERY, MICHAEL, *A Field Guide to the Insects of Britain and Europe* (2nd edition), William Collins, London, 1976

CHINERY, MICHAEL, *Spiders*, Whittet Books, London, 1993

CLAUSEN, L., *Insect Fact and Folklore*, Collier Books, New York, 1954

COWAN, F., *Curious Facts in the History of Insects; Including Spiders and Scorpions*, J. B. Lippincott, Philadelphia, 1865

DARWIN, CHARLES, *The Voyage of the Beagle*, distributed by Heron Books, 1968

DUNCAN, JAMES, *Beetles, British and Foreign*, London, 1880s (originally published as *The Natural History of Beetles* in 1835)

FORD, E. B., *Butterflies*, Collins Clear-Type Press, London and Glasgow, 1971

GILBERT, P., *Butterfly Collectors and Painters. Four Centuries of Colour Plates From the Library Collections of the Natural History Museum*, Beaumont Publishing, Singapore, 2000

GULLAN, P. & CRANSTON, P., *The Insects: An Outline of Entomology* (2nd edition), Blackwell Science, Oxford, 2000

HARRIS, M., *The Aurelian or, Natural History of English Insects; namely, Moths and Butterflies : together with the Plants on which they feed, etc.*, printed for the Author, London, 1766

HOWARD, L. O., *History of Applied Entomology (somewhat anecdotal)*, volume 84, Lord Baltimore Press, Baltimore, 1930

NEWMAN, E., *An Illustrated Natural History of British Butterflies*, William Tweedie, London, 1871

NEWMAN, L. HUGH, *Looking at Butterflies*, William Collins, London, 1977

OSBORN, H, *A Brief History of Entomology*, Spahr and Glenn Company, Columbus, Ohio, 1952

SALMON, MICHAEL A., *The Aurelian Legacy: British Butterflies and Their Collectors*, University of California Press, Berkeley and Los Angeles, 2000

SMITH, R.F.; MITTLER, T.E.; & SMITH, C.N., *History of Entomology*, Annual Reviews Inc., California, 1973

SPEIGHT, M.R.; HUNTER, M.; & WATT, A.D., *Ecology of Insects*, Blackwell Science, Oxford, 1999

STANEK, V. J., *The Illustrated Encyclopedia of Butterflies and Moths*, Octopus Books, London, 1977

TOPSEL, E., *The history of four-footed beasts and serpents. Whereunto is now added, The theater of insects /collected out of the writings of C. Gesner and other authors by E. Topsel*; by T. Muffet, London, 1658

ARTICLES

BON, MONSIEUR, "A discourse upon the usefulness of the silk of spiders. By Monsieur Bon, President of the Court of Accounts, Aydes and Finances, and President of the Royal Society of Sciences at Montpellier. Communicated by the Author," *Philosophical Transactions*, 1710–12, 27: 2–16

COCKERELL, T. D. A., "Dru Drury, An Eighteenth Century Entomologist," *Scientific Monthly*, 1922, 14(1): 67–82

ESSIG, E. O., "The Value of Insects to the Californian Indians," *Scientific Monthly*, 1934, 38(2): 181–6

Hogue, C., "Cultural Entomology," *Annual Review of Entomology*, 1987, 32:181–99

Weiss, H. B., *Thomas Moffett, Elizabethan Physician and Entomologist, Scientific Monthly*, 1927, 24 (6): 559–66

USEFUL WEB SITES

American Board of Forensic Entomology, *Official Page Of The Board*: www.missouri. edu/~agwww/entomology/

American Museum of Natural History, New York: www. amnh.org/

Colorado State University, *Entomology*, Department of Bioagricultural Sciences and Pest Management: www. colostate.edu/Depts/ Entomology/ent.html

Gordon John Larkman Ramel, *The Earthlife Web (insects)*: www.earthlife.net/insects/

Meyer, J.R., *Compendium of Hexapod Classes and Orders*, Department of Entomology, College of Agriculture and Life Sciences, North Carolina State University: www.cals.ncsu. edu:8050/course/ent425/ compendium/index.html

Miller, G & Peterson, R., *Insects, Disease and History*, Montana State University: http://scarab. msu.montana.edu/historybug/

National Museum of Natural History, Smithsonian Institute, Washington DC: www.mnh.si.edu/

Natural History Museum, London: www.nhm.ac.uk/

Sear, D., *Bugbios*: www. bugbios.com/

University of Nebraska, *Nebraska Entweb*, Department of Entomology, University of Nebraska Lincoln: http://entomology.unl.edu/ index.htm

VanDyk, J.K. & Bjostad, L.B., *Entomology Index of Internet Resources*, Iowa State University: www.ent.iastate.edu/List/

Vanuytven, H., *Arachnology: The Arachnology Home page*: www.ufsia.ac.be/Arachnology/ Arachnology.html

CLUBS AND SOCIETIES.

Amateur Entomologists' Society, PO Box 8774, London SW7 5ZG, UK
E-mail: aes@theaes.org
www.theaes. org/

Australian Entomological Society
www.agric.nsw.gov.au/Hort/ascu /myrmecia/society.htm
#Membership details

Entomological Society of America, 9301 Annapolis Road, Lanham, MD 20706-3115, USA

E-mail: esa@entsoc.org
www.entsoc. org/about_esa/

Entomological Society of Canada, Head Office, 393 Winston Avenue, Ottawa, Ontario, Canada K2A 1Y8
E-mail: entsoc.can@sympatico.ca

Entomological Society of China, 19 Zhong Guan Cun Lu, Haidian Beijing, China
E-mail: zss@panda.ioz.ac.cn;
http://panda.ioz.ac.cn/ ioz/esc.html

Linnean Society of London, Burlington House, Piccadilly, London, W1J 0BF, UK
www.linnean. org/index.htm

Linnaean Society of New York, 15 West 77th Street, New York, NY 10024, USA
http://linnaeansociety. org/

Royal Entomological Society of London, 41 Queen's Gate, London SW7 5HR, UK
E-mail: reg@royensoc.co.uk
www.royensoc.co.uk/

INDEX 1

INDEX 2

ACKNOWLEDGMENTS

We thank Caroline Holebrook and David Bedford at Canterbury Christ Church University College library for their cheerful sufferance of our many requests; likewise the librarians in the Entomology and Botany libraries of the Natural History Museum, London, and in the Royal Entomological Society of London library, in particular Berit Pederson. We also thank John Badmin for the loan of *The Naturalist in Britain* by David Elliston Allen; Jane Seaman for Charles Darwin's *The Voyage of the Beagle;* and Mildred Warner for Mrs. Beeton's *Book of Household Management.*

Special thanks go to Ian Haigh and Vivienne Ponsonby for their support and invaluable criticisms of the manuscript. Thanks also to Caroline and Hayley Ponsonby for their helpful suggestions and delightful distractions. We also thank Dr. Kevin Carlton of Canterbury Christ Church University College for relieving some of the academic load during the preparation period, and likewise Cynthia and Keith Beverley, Dr. Georges Dussart, Peter Gilchrist, and friends for their enthusiastic interest and advice.